FRANK I

REAL COPS

"One of the best books I have read in a long time. Real Cops kept me glued until the end."

~ Eduardo Jordan,
(Former Special Agent, U.S. Department of Justice)

"I liked Lee Childress, a new agent, a virginal, straight-laced guy, dealing with gangsters and hookers. Meanwhile, H.T. 'Honey' Ross, a cynical hair-bag veteran agent, shows him the ropes. The depiction of crooks like Baby Face Nelson ring true and accurate."

~ Bob McVeety,
(Former Manhattan DA Investigator/Police Officer)

DANCING MAX HITS GUADALCANAL OR WHEN IN DOUBT, RHUMBA

"Frank Hickey's stories remind you of Elmore Leonard, Denis Johnson, Pete Dexter or any other great storyteller writing about the underbelly of society. His characters and plots are unique; Frank's life experience bleeds through onto the page. He's authentic as hell!"

-Andy Rosenzweig, (Former Chief Investigator, Manhattan District Attorney's Office)

FUNNY BUNNY HUNTS THE HORN BUG

"It's a great read. There is nothing fake about this book. It is authentic. The characters seem like real people that you'll never forget. The NYPD politics ring true, just as insane as when I was there. This book deserves to be read and re-read and studied as criminal justice literature."

~ Daniel Vona,
(Deputy Commissioner, NYPD [Ret.])

Ukraine, Whaddya Gonna Do?
or
Sunshine, Salo and Sirens

By the Same Author

Books

Real Cops
The Gypsy Twist
Funny Bunny Hunts the Horn Bug
Brownstone Kidnap Crackup
Can Showbizzers Crush Crime?
Softening Flatbush
When the Whistle Blows, Everyone Goes
Max Wisecracks Hollywood or Foxtrotting for Justice
Love Finds Max Royster or Kissing in the Slush after Sixty
Dancing Max Hits Guadalcanal or When In Doubt, Rhumba

Feature Films

Spy, The Movie
(co-written with Charles Messina & Lynwood Shiva Sawyer)

Ukraine, Whaddya Gonna Do?
or
Sunshine, Salo and Sirens

by Frank Hickey

Brooklyn • London
Fincastle

Copyright © 2025 Frank Hickey

Original cover art by Nad Wolinksa (All rights reserved)
Cover Titling by Lynwood Sawyer
Book Design by D. Bass

ISBN: 978-1-7331750-8-1

All rights reserved. No part of this book may be reproduced or utilized in any form or by any means, mechanical or electronic, including manual re-input, photocopying, scanning, optical character recognition, recording or by any information storage and retrieval system without permission in writing from the copyright holder.

For further information, please contact:
http://frankhickey.net

Published by Pigtown Books

Pigtown Books Logo and Colophon designed by Richard Amari

Dedicated to
all the Ukrainians everywhere
who suffer from this war.

Salo is a traditional Ukrainian food, made of cured pork fatback or boneless pork belly, usually without lean meat and skin. It has become an honored national symbol and the subject of countless recipes and proverbs. For example:
   "If I were a lord, I would eat salo with salo."

Contents

1	Why Ship Out Overseas?	1
2	Rockets Wake Me	5
3	Rocket Day Dawning	13
4	And Just How Was Your First Week Here?"	15
5	Cussin' in Kherson	19
6	Patty Duke is 'Billie	21
7	Teacher Spanks Me, Again	23
8	Lovely Friday	27
9	Sometimes You Feel Guilty	29
10	Hemingway Strips	31
11	Rocket Shelter Dancing	33
12	Lying in the Shelter	35
13	'Star Wars' Scares Me	39
14	And Now for Something Completely Different	43
15	Nightmare	45
16	Saturday Night in Old Ukraine	47
17	Would You Feel?	49
18	What Would You Say in the Shelter?	51
19	Ukrainian Battered Women	55

20	Kirsti, Why Don't You Make Me Real?	59
21	Black Sea Beach	61
22	Rockets Hit	63
23	Help Us If You Can	65
24	Don't You Know There's a War On?	67
25	Sweaty Fruit	69
26	Changing Isolation	71
27	How it All Began, New Mexico Cowboy Detective to Ukraine	75
28	Breaking Out	79
29	Colorful New Mexico	81
30	Can You Pronounce the Name Pryzemysl?	83
31	Will Russia Attack Me, Me, Who Everyone Loves?	85
32	Strong Forts	87
33	Poland	89
34	Sky Pilot Fights Back	91
35	Walking Against the Traffic	93
36	Singing Opera During Air-raids	97
37	Teaching What? What?	99
38	Odessa Da-Da Sky Pilots	101
39	"You Think You Know Your Neighborhood."	103

40	Policing Meets Young Ukraine	105
41	'Well, my Mind is Going Through Some Changes.' (Buddy Miles, musician, 1971.)	107
42	"Bombings Morph to Backups," I Say	109
43	DREAMS	111
44	During Air-raids, Playing 'the Blue Danube.'	113
45	Wedding Chit-Chat	115
46	Bread Line, Water Line, Riding Circuit	117
47	Ukrainian War Warp Speed-Dating	119
48	On the Road Again	121
49	Horse Sense	123
50	Fighting Women's Run Fu	125
51	Odessa August Routine Days	127
52	Ukraine August Songs Cheering Our Hero Up	129
53	Fear Too Much	131
54	Critics Boo My Notes or What Would Mother Hickey Say?	133
55	Chance Conversations	135
56	Ukrainian Cuisine	137
57	Suspicion Torments my Heart	141
58	Surrender? Like Holland and France?	143

59	Polly-Wolly Polly-Tics	145
60	What Would You Buy for Free?	149
61	Dvorak's New World Symphony	151
62	Sunset Dancing on the Black Sea	155
63	Ambassadors to Normal	157
64	"Something Medical Hit Me on Saturday Night."	159
65	Yesterday Odessa's Humidity Hit 85 Percent.	161
66	Odessa Traffic	163
67	Talking through the Goat Rodeo	167
68	Exploding Head Syndrome	169
69	Looking Good, Mark Hamill	171
70	Big Apple Prep	173
71	If You Love Somebody....	175
72	A Real Odessa Guy	177
73	Why Is?	179
74	As a Kid, I Wanted to Be European	181
75	Kirsti, I Think.	183
	About the Author	185

1

Why Ship Out Overseas?

"Hello," I say to the Ukrainian Consulate clerk in New York. "I want to help war refugees as a volunteer."

His face washes blank.

Behind his bulletproof glass, he mumbles something.

"HUH?" I utter.

"Don't know," he says.

Ex-detectives like me are used to this kind of government cow-flop. I feel silly.

He makes me feel silly.

"What volunteer groups are helping Ukrainian refugees?" I ask again.

"Dunno," he repeats.

Ukrainian folks line up behind me. They cough and fidget. They want their turn at this information wizard.

A thought seems to strike him.

"Try, Red Cross," he wheezes.

"That's like telling me to write to the White House," I say.

"Don't know, this," he says.

He shrugs.

In teen years, I started as an Angry Young Reporter. It starts coming back to me.

IT'S YOUR COUNTRY! I want to shout.

THIS WAR STARTED SEVEN MONTHS AGO!

TRUMP CALLED THIS INVASION 'WONDERFUL.'
HOW MANY AMERICANS AGREE WITH THAT?

150,000 Ukrainian-Americans lived in New York City before the invasion.

"Here," I say to him. "My application to join your Ukrainian Foreign Defense Legion. With copies of my passport and driver's license, as requested."

"Not here," he says. "Military Attaché to United Nations. 220 East 51st street. I very busy."

"Gotcha," I say. "I'm no fearless Rambo warrior. But I need to join up."

He gives me the Postal Clerk Glare.

Like most government workers at a window, he wants one thing. He wants you to *Go Away.*

Excited, I leave, bring my application to the Military Attaché and snag him leaving the office. Tall, rangy, crew-cut and in green army colors, he is courtly and soft-spoken.

He takes time to scan my resume.

"You are Ukranian?" he asks.

"No, sir. Irish and Jewish."

"You have detective experience," he says.

"Since 1975," I answer.

"Would you want to investigate war crimes?"

"Yessir!"

That came out fast.

"Okay," he says. "We see."

For weeks, I return and make cold calls on him. Real detectives never call and warn the target beforehand.

He is always out. Emails and phone calls go lost into cyberspace.

The friendly Ukrainian soldier guarding the front door gets used to seeing me.

"Maybe you go," he suggests.

"You're right," I say. "Maybe I go."

2

Rockets Wake Me

This is my first night rocket attack.
It seems weird.
I feel stupid.
"Russians are launching rockets at Odessa," I mutter. "Near my hotel. Near ME. Whom everybody loves."
Noise sounds closer.
I cannot figure out their direction.
BOOMP! BOOMP!
Then quiet.
Another one hits.
They all sound close.
This seems like waiting for rain to stop. There is no logic to it.
"This changes the game," I mutter. "How much good am I really doing here, working with refugees? Unloading trucks, making camouflage nets, teaching English to Starbucks teenagers."
My phone blows a siren.
"Attention!" a man's voice comes from the phone in English. "Air-alert! Go to the nearest shelter. Don't delay. Your over-confidence is your weakness."
The Russians launch rockets from ships on the Black Sea two miles away. Their rockets can soar unseen through the thick black night and destroy everything inside a half mile area.

Ukraine, Whaddya Gonna Do? or Sunshine, Salo and Sirens

Rolling out of my hotel bed, I slip on my shorts. Fingers twist the belt. I cannot untangle it.

I keep trying. On the police handgun range, I used to have this same dexterity problem.

CRUMP!

Noise feels like the horror surprise of a car crash.

It is embarrassing because I don't expect it.

I can't expect it.

What do you wear during a bombing?

I go for the casual look.

Odessa is a beach town, after all.

The night is cool. This is my chance to wear the red T-shirt. In this worldwide heat wave, this shirt is too hot for the daytime.

Rescue crews can see my red shirt in bomb rubble and find me that way.

Noise thunders as I tear through wardrobe planning.

On their balconies outside, other hotel guests mutter. Cigarette smoke smells.

I sling on my blue Canterbury prep-school knapsack with the school logo on it. It is my best knapsack, well-made and sturdy. On the front, I had written the words 'Tuition Revenge, 1967 - 1971.' This is preppy humor.

Holden Caulfield, where are you tonight?

Inside the knapsack lie my first-aid kit, flashlight, bottled water, popcorn and granola. I hate granola.

But Odessa stores lack variety for emergency food.

The elevator is dead. Did a bomb hit it?

On my right, the stairs bloom dark.

Maybe some shrapnel nailed the electricity.

Limping down three flights to the lobby, I find four

lanky teenagers smoking and staring at their cellphones.

Grusa, a wrinkled woman housekeeper with a reddish drinker's face sports a violet T-shirt. With English letters, her shirt reads 'the Future is Romantic.'

Nobody talks much.

Maybe I was expecting a speech?

If they notice me at all, they see a 70 year old American, of Irish and Jewish blood. I'm not Ukranian. I sport two knee braces, cowlicky brown hair with sideburns turning gray and nearsighted blue eyes blinking in the dark.

A tin Medic-alert bracelet on my wrist warns of heart trouble. After the heart operation, my energy comes and goes.

Tonight, it is going. Doctors warned me to stay out of this war zone, taking four pills a day.

Rodeo work is dangerous. When a steer tosses a cowboy onto the ground, that steer can easily stomp, gore and kill the cowboy. So rodeo clowns dance out in front of the bull to distract him. The clowns are brave cowboys who grew up around steers.

They cavort, do pratfalls and annoy the steer until the cowboy leaps to safety.

Sometimes I feel like a rodeo clown in Ukraine, trying to distract Ukrainians from this horror and cheer them up.

The lobby windows look outside. They show Odessa's stately gray stone buildings in French and Italian classical style, framed by large trees. An ambulance roars past on the flagstone street, blue emergency light winking.

The youngsters slump downstairs to the basement bomb shelter, faces almost kissing their phones. I follow them.

Our shelter room is dark and smells like a baboon's sock.

The toilet is a place to avoid. The water stopped flowing days ago.

This room was designed for four people.

12 of us crowd inside it.

"The elevator seems unwell?" I ask the desk clerk, Daria. Speaking adequate English, she looks about 25 and scared, with long dark hair and pale skin.

"It just stopped tonight," she says in her soft accent. My two bad knees hear her. They remember the stairs and start aching ahead of schedule. "Before the bombing."

"Glad I did my shopping yesterday," I say. I try to sound like the wisecracking New York detective that I used to be.

Nobody reacts to that, either.

BOOM!

Another one hits nearby.

I imagine the building swaying and collapsing around me. Years ago, California earthquakes had given me that same sick roller-coaster feeling. These rockets will not warn us before they hit.

"How can you be so calm?" Daria asks me.

Daria is wrong. I'm not calm. I'm fighting to look normal. But my insides are pounding like a runaway horse.

"Your website says you are a writer and policeman," Daria says. "Maybe because you were a policeman, you are not afraid?"

"Interesting question," I say, stalling for time. "In America, a policeman on duty dies every three days. I wore blue for 16 years and I was afraid all the time. But I learned to look at statistics. Can you look at tonight mathematically?"

"Mathematically," she echoes.

"How many people live in Odessa?" I ask.

"A million," she says.

"Out of that million, what are the chances of a rocket hitting us here?"

I let that sink in. Then, I moved for more distraction.

"Friday night, I went to my ballroom dancing place," I say.

"Why do you go to dance?" Daria asks. "For girls?"

"I love dancing," I say. "Danced around the world. Mongolia, Borneo, India."

"Friday, was more bombing?" she asks. "Tell me, please."

My distraction plan is not working.

"Near my dance place," I say, "a rocket blew up an entire block. The shock waves broke the windows and spread glass all over the dance floor. I saw families standing outside the twisted steel and bricks. Yesterday, that mess was their home."

"Do Americans know how bad is this war?"

"We should," I say. "We suffered a Vietnam War for 15 years and lost 58,000 dead. Your war is just 500 days old and you have 80,000 soldiers killed."

"This bombed building is near the port?"

"You know this, huh?" I say. "Yep. The Russians want to destroy all the grain in Odessa's port. That grain feeds hungry families around the world, in Africa and Asia. Thinking about it makes me hungry."

Opening up my bag, I pass my popcorn bag around. Nobody takes any. Ukrainians are a proud people. Trying to look carefree, I munch on the popcorn kernels. I want the others to remember happy times chewing popcorn at the movies months ago, before the invasion.

"You hungry?" Grusa of the romantic violet T-shirt asks me in broken English. The half-light catches her brown eyes on me. Ukranian women have eyes like jewels.

"Grusa, I gotta work in a few hours," I say. "Air-alerts always make me hungry. Maybe it's the stress."

"DEPRESSIA," she says in Russian.

"I can figure that word out on my own," I say. "Ukraine suffers the world's highest rate of mental illness and depression in the world. We've gotta fight it as best we can."

"So you dance?"

"That's right," I say. "That's why I danced with my friend here in the shelter at our last air-alert. Handle that old DEPRESSIA."

Everyone waits.

Bombing stands everything on its head.

Still chewing popcorn, I wonder about the young mothers next to me with kids, all shook awake like I was. How tired will they feel at their jobs, around three in the afternoon?

"They bomb us last night, too," Daria mutters.

No more rockets hit.

"Attention!" my phone says in English.

"The air-alert is over. May the Force be with you."

Everyone in the shelter breathes out.

Youngsters shuffle to the door. The mothers group their kids.

"I think you can go," Daria says.

Maybe I am being dismissed.

Padding upstairs again, my body wants more bed.

In my room, I open a bag of onion crackers and wash them down with sunflower oil.

"Comfort food," I murmur.

I lie on my bed.

My phone buzzes.

"Attention!" the same voice on my phone says. "Air-alert. Go to the nearest shelter."

3

Rocket Day Dawning

Rocket day dawning in Odessa has me stumbling tired on the tree-shaded streets.

"This attack was the closest one to here," a woman says in English outside my cafe door. "And they will attack us again tonight."

"Hope not," I say, scared. "Lost three hours sleep. I'm dragging tail."

I hear myself trying to sound tough. Night will come too quickly. That scares me.

My body feels like it used to, going to court after street-policing in the Los Angeles 'hood all night.

The Russians might rocket us again for a week.

Last night feels primitive, medieval.

"The news says that the rockets did not hit anything," I say. "But something made a lotta noise near my pillow."

"Putin does this to destroy Odessa's port on the Black Sea," she says. "Stop our grain. World is hungry."

This afternoon, a long white stretch limousine flashes past on the boulevard. A blonde bride in white dress and veil beams, standing up in the sunroof, coupled to the groom. They both look about twenty.

Probably Odessa has learned how to live with rocket attacks. Not me, not yet.

Ukraine, Whaddya Gonna Do? or Sunshine, Salo and Sirens

At work, everyone seems more tired than usual. We cut camouflage rags slower. The coffee machine putt-putts more.

We down coffees. Fingers paw for sugar cookies.

Noise hits. I flinch.

The sun sets. Outside my window, the alley pencils gray getting darker.

4

And Just How Was Your First Week Here?"

"How was your first week here in Odessa?" Kirsti asks me on Monday at our Gorky Park cafe.

"A goat rodeo," I say. "Would you try going to a war zone where you don't speak the language? And ask for volunteer work?

The Ukranian Army keeps rejecting me.

I'm no fearless warrior. A carload of things scare me. But I gotta do something.

"I bounced from one charity group to another. Big names. International names. Spoke nice to some English speakers in these groups.

"They said they would contact me soon about volunteer gigs. Like a detective canvassing a crime scene, I spoke to everyone. Local church. Organist. Stopped a holy man in his habit on the street. Still don't know his religious affiliation. Coulda been High Priest Anton LaVey in the Church of Satan.

"Saw parked volunteer jeeps and put my contact cards under their windshields.

"Heard nothing," I gripe.

"Maybe you scared people," Kirsti says.

"Compared to the Russians? Since the war started, Russians have killed hundreds of Odessans. They mined the harbor. May come ashore here any day in amphibious assault."

"My father says that charities are frauds," Kirsti says.

"I'm straining not to get cynical," I say.

"As a detective, I took down two fake police charities. Maybe charities are just slow and clannish, like most private companies. I'm still finding out.

"I like Odessa. The buildings are elegant classic stone in the French or Italian design. They don't resemble the ugly Soviet style. Every shade tree stands at least 30 metres feet high, over gray flagstone streets.

"South of downtown, apartment buildings lie quiet in huge green spaces of trees, grass and gardens. The Black Sea beach sports lively loud swimsuit and bikini swimmers, trying to shout down the war."

"Did you find a hotel?" Kirsti asks.

"Yep. My three rules. Budget, no drunken discrepants in the hall and walk to whatever volunteer work you score. After three days of disappointment from getting no work, I knocked on a nearby relief center door. Said that I wanted to help war refugees.

"Give them credit, they put me to work right away. No email delays, no get-aquainted zoom calls next month. Unloading trucks, hauling garbage and cutting cloth rags for camouflage nets."

"I did that," she says. "Boring."

"We agree. But volunteer work is like old-time vaudeville. When you find a show that suits you, stick with it. Volunteer vaudeville.

"Third day, boss man says in his cracked English, 'You should see things in Kherson. Go with my friend tomorrow.'

"Kherson is near the front line. So, I agreed. That evening, I went to the park and a supermarket. Somebody picked my pocket and stole my phone.

"A Suspicious American Savant asked me later, 'Did the secret police there seize your phone because of what you wrote about them?' Penetrating question.

"I couldn't buy a new phone. The phone stores close early in Odessa. And no dealers speak English well. Sure, I knew the dangers of heading into a war zone with no phone but I wanted to go there. This might be my only chance."

"You're stupid to risk that," Kirsti says.

"Maybe," I say. "The next day, we drove east. Just twenty miles from Odessa, I saw buildings destroyed by Russian rockets. The Russians had tried to cut Odessa off from the rest of Ukraine. More and more wreckage followed.

"In Kherson, whole blocks were rubble. The driver dropped me at a volunteer center. 'I come get you, two, three hours,' he says in okay English.

"I work that afternoon with other volunteers. More medical supplies and food to load into trucks. The volunteers wear military uniforms and weapons.

"'Can you support us with money?' a shaved head, tattooed volunteer asked me on Google Translate. I noticed the combat knife on his belt. Since the invasion, Ukraine scrapped all its gun and knife laws.

"I pleaded the Penniless American Volunteer. He persisted. Everyone watched this.

"After a suitable show of fake bravery, I took a night walk outside. The building across the street had been hit by rockets. As I walked, noises clicked from the ruined building. It sounded like a woodpecker. Bricks were still falling from the walls. They made that noise. Some shattered on the sidewalk.

"Then artillery sounded. Pretty close. Too damn close."

"Why didn't you leave?" Kirsti asks.

Ukraine, Whaddya Gonna Do? or Sunshine, Salo and Sirens

"How? Hitchhike at night? And go where? It's a two-hour drive to Odessa.

"I tiptoe into an empty room in my volunteer house, pile furniture against the door and try to sleep. Nightmares come. I wake up and fidget. Someone pushes against the door. My furniture falls."

5

Cussin' in Kherson

"The furniture fell onto my bedroom floor," I tell Kirsti. "I'm among armed strangers asking for money at night. They're trying to get into my room. This scares me.

"But the door held. I remember that nightcrawlers do not like noise.

"So, I bellowed a kinda Brooklyn/Ukranian bellow.

"Something along the lines of '"HEYOUSE DERE, BABY, DON'T CHA KNOW, TRYNA SLEEP HERE?'

"The nightcrawler stole back into the night.

"More artillery shells THUMPED! nearby.

"Over morning coffee, I approached one of the more amiable volunteers and spoke in baby-talk Russian.

"'Say, old buddy,'" I began. 'Might you maybe know when my friend comes? To carry me back near Odessa.'

"He looked unconcerned.

"In a word, insouciant.

"Striving for calm, I repeated the question.

"'Who are you?" he asked in Russian.

"That question echoed," I tell Kirsti.

"I had no quick glib answer.

"Kirsti, the morning degenerated into a hair-ball of confusion. Nobody knew nothing."

"Serves you right, going there," Kirsti snaps.

"Finally, after much pleading and mayhap groveling," I say. "I talked my way into a fast ride to Odessa. The ride dropped me somewhere lost in Odessa. But, by dead reckoning, I found my hotel's neighborhood.

"Whenever a lot of raggedy Odessans are chatting with themselves, I know that I am near my hotel."

"That's because your hotel is raggedy cheap," she says.

"Hey," I say. "Remember that I went to the Ukranian Foreign Legion to join up. Six times. They call me too old for their service. I bet that their bed-buggy army bunks are a lot worse than my hotel."

"You can't join the Legion at your age!" she snaps. "You're just too old. Whenever you move, you creak. When are you going to grow up?"

6

Patty Duke is 'Billie

"Does anyone remember the 1965 Patty Duke film 'Billie'?" I ask Kirsti.

"Should we?" she asks. "I wasn't born yet. Why do you ask?"

"I use it to teach at the Ukranian Domestic Violence center," I say. "My women students always want happy films. Between the war and their own troubles, they want something that makes them laugh."

"Not just laugh at you?"

"They do that, too," I agree. "In the film, Patty Duke plays Billie, a tomboy who can outrun every boy on her high school track team. It's a lightweight musical comedy with good song and jazz dance numbers. The songs like 'A Girl Is a Girl Is a Girl!' stay in your head."

"Maybe in your head," she says.

"You should see this film running free on YouTube," I say. "The dances were done by that famous man Gower Champion who won 8 Tony awards.

"And the writer, Ronald Alexander, was primarily a playwright. His plays did well, running on Broadway for years.

"Audiences liked his wit. So do I. He wrote a satire about the TV industry, entitled 'Nobody Loves an Albatross.'

"Why do you waste their time with this lightweight junk?" she probes.

"Because it's early Feminism," I opine. "Experts say that women with a strong self-image will take action against an abuser. They're more likely to call 911, move out of the relationship or seek counseling."

"So, this Ronald Whoever had to start somewhere, with this silly movie?" she asks.

"Yes," I answer. "He began life as a boxer and a jazz singer. To get an easier existence, he started writing. He did well."

"And now he's an unknown writer?" she smiles. "Kind of like you are?"

"You're reading it right," I say. "Like all unknown writers, he had to Start Somewhere."

"And that's what you're doing, writing here in Ukraine," she says. "At 70? Starting somewhere."

7

Teacher Spanks Me, Again

"Back in New York, 1988, I'm watching Joey T.," I say. "When Joey T. leaves his home, I run up, grab him and holler 'Police, Joey! You're under arrest for fraud.'"

"My hero," Kirsti smirks. "This was when you were a detective?"

"A naive, ignorant and innocent detective," I answer. "Joey T. had run a fake charity for 12 years and cheated donors out of millions of dollars."

We are lying on Odessa's Black Sea beach.

Waves break low and smooth, as usual. Bluish water crashes cream against the sand.

Kirsti sounds cranky today. Russian rockets hit Odessa last night. People died. Maybe she feels the strain.

"That was the first charity I ever investigated," I say. "Maybe it soured me on charities."

"Was he helping anyone?" Kirsti asked.

"Sure," I say. "Himself."

"What was his fake charity?"

"Kirsti," I say, "never ask a gently aging detective like me about his old cases. I'll blab until sunrise.

"For 12 years, Joey T. ran a false charity outfit. He claimed that the outfit helped police widows and orphans. He hired 75 con men, immature guys who needed to pose as cops and real-life killers.

"They solicited and pressured merchants to donate. On the average, Joey pulled in $600,000 a year, in 1987 dollars."

"And you fought against him for the good of mankind," she says. "And now you don't trust charities. Because they don't immediately offer you splendid wonderful volunteer work."

"I trust some charities," I say. "I'm just trying to volunteer for anyone helping Ukranian refugees. And asking for work but not getting any is frustrating."

"But you're already doing some volunteer work," she says.

"Yeah. But I wanna do more. It's like policing. Sometimes, you're not sure how much good you're doing."

"You're writing about this?" she asks.

"Of course I am. But are you reading my sketches?"

She smiles a lazy beach smile.

"Doesn't every writer ask that question?" she asks.

"Sure they do," I said. "But I'm asking it. ME."

"You call yourself 'ignorant,'" she says. "You still are ignorant. Why should you be able to volunteer easily during this war?"

"Because I want to," I say.

"How much does that matter?" she asks. "What you want. We Ukrainians WANT to survive rocket attacks every night. Forever. You'll be going back to New York. We cannot.

"Do you know the logistical problems in having volunteers for a charity? Someone has to train and supervise them. Or else, you have chaos.

"And what will you do, in this volunteer work?"

Her sharp tone makes my breath stop short. My stomach tightens on the sand. It feels like Teacher is spanking me. Again.

"I'll do anything," I say. "Anything not illegal or fattening."

"You see?" she asks. "You don't even know what you'll be doing. You don't speak Ukranian. And your trying to speak Russian gives me the giggles. And how long will you stay in this imaginary volunteer agency?"

"That depends on the work," I say. "If I think that they are really helping refugees."

"Oh, wonderful," she answers. "Remember I was a business major. That's not what every supervisor wants to hear. That you'll stay as long as you approve of his work. Drift in and move on, like a 70-year-old beer-belly Huckleberry Finn."

"That's not what I mean," I say.

"And this boss can't control you with money because you're a volunteer," she says. "An enigma. You're not even Ukranian. Everyone calls you a mystery. And a supervisor's nightmare. If I were running a charity, I would not let you enter my doorway."

8

Lovely Friday

"It's finally happening," I tell Kirsti.

"What?" she asks. "You're modifying your gender?"

"No," I say. "Forgetfulness."

"Senility," she snaps.

We sit in our sunny and windy park.

"I can't remember how many times we had air-raids last night," I say. "Or how many times I hit the shelter."

"Dementia," she continues. "Remember, you're a lot older than I-"

The air-raid siren blows.

"Attention!" my phone raps out. "Air-alert!

"Go to the nearest shelter. Don't delay. Your over-confidence is your weakness."

"Mark Hamill and his air-alert announcement can go sandpaper a monkey," I say. "Today's a lovely Friday. April is just twenty days old today and I want-"

BOOM!

Something explodes nearby. A shock wave pushes my hair back.

It reminds me of shooting at the Los Angeles police shotgun range. Back then, the muzzle blast pushed my hair, in the same scary way.

BOOM!

This one feels closer.

"Where you going?" she asks.

"To the bomb-shelter," I say. "It's not a lovely Friday anymore."

9

Sometimes You Feel Guilty

"Sometimes, you feel guilty, enjoying Odessa," I tell Kirsti.

At sundown, we walk through light falling snow near the Opera House.

"Why?" she asks.

"Because it's war," I say. "Children are dying three hours drive from here. America is sending nothing to help. And we're styling out to watch ballet in thrift-shop clothes. I need to do more."

"Maybe your clothes are thrift-shop," she says. "Mine are not."

"My friends back home may think that I'm young Hemingway," I say. "Driving his ambulance as shells burst around me. Or chasing Putin across midnight rooftops, with my gold-and-blue detective shield swung around my neck and a black Glock gun in my mitt."

"You have an active fantasy life," she says. "I'm reading your Guadalcanal detective book."

"Slowly?" I ask. "Savoring every word? Back to the war. Seven times, I made cold calls, trying to join the Ukranian army. Once in the army, I might be freezing and cleaning toilets as the Russians launched rockets at my aging body. Do you think that the army was going to send me to the ballet?"

"Do you WANT to clean toilets?" she asks.

Ukraine, Whaddya Gonna Do? or Sunshine, Salo and Sirens

"Not partic-u-larly," I answer. "But as Daddy Hickey used to say, it's steady work. And anything that helps Ukraine, that's what I want to do."

An air-alert siren blares on the corner. Nobody reacts to it.

"And more Russian rockets are hitting Ukraine this week," I say. "Because Ukraine is running out of defense missiles to protect us. That's your ribcage and mine, Kirsti. Recent aid dropped by almost 90 percent. American help is ending soon."

10

Hemingway Strips

"Something wakes me here in Odessa this morning at three," I tell Kirsti.

She nods, sipping her coffee.

"Kinda weak whitish lightning lights the sky from my hotel window," I go on.

During her morning coffee, Kirsti is a good listener.

"I wonder if it's artillery fire, far away. Something out of a Hemingway story. It's the romantic in me."

"Then my phone app comes alive. 'Air alert,' it says in a flattish Midwestern accent. 'Go to the nearest shelter. Don't delay. Your over-confidence is your weakness.'

"Our local siren wails.

"I move fast and throw on shorts with a red T-shirt. Rescue crews can see red better than any other color. Heavy socks and work boots for moving safely through wreckage all slip onto my feet. My night pack with First-Aid kit, granola, popcorn, phone charger and notebooks follow. Fumbling, I thumb one contact lens into my left eye. Rambo never has to do that."

"Your hotel clerk must think you are crazy, carrying all that," Kirsti says.

"That's me. I wanna be crazy and alive. Not logical and crippled, with a belly bag. You realize that I can't get any kind of medical insurance? No insurance company will cover anyone in Ukraine.

"So I'm out the door in under four minutes and step downstairs to the lobby. Our shelter is near the lobby. In yesterday's air raid, guests and workers filled the lobby. This morning, the lobby stands empty and dark.

"Maybe the others are ignoring the siren. Maybe their over-confidence is their weakness.

"I sit and write this. I feel alone and silly for reacting to the alert. My phone sounds. 'The alert is over,' the same voice says. 'May the force be with you.'

"That's the catch-phrase from the film 'Star Wars' in 1977. That year seems like a simple and innocent time.

"I go upstairs, strip and try to sleep."

11

Rocket Shelter Dancing

"Why are we dancing in the shelter during this air-raid?" Kirsti asks.

"Trying to fight my fear," I say. "For real.

"This makes four air-raids today. I was just starting to write about our last alert. Then this one sounded."

"What's this song?" she asks.

"'Perfidia' by Glenn Miller," I say. "From 1941."

"Who is this Miller?" she says.

"When World War Two erupted, Miller reigned as America's favorite musician," I answer. "He tried to join the military and sacrifice a salary of $330,000 in today's money. The Navy rejected him because he was 38, a father with kids and wore eyeglasses. But the Army took him, trained him and ranked him as a major. The world was on fire then, much worse than now. Miller wanted to do something. Anything."

"Like you," Kirsti says.

"Nothing like me," I say. "I'm no hero.

"Miller did two years of active military duty overseas. He died when his plane went lost in the English Channel. They never found the wreck.

"By dancing to his songs, I remember him and learn how to get through this war."

We keep dancing.

In the shelter, some visitors ignore us dancing.

Their faces bend towards the small gray rectangles of their cell phones.

"What else do you learn?" Kirsti asks.

"I learn that Ukrainians think that all America supports them," I say. "My students think so. Two high Ukrainian police officials thought so."

"Is it true?"

"I'm afraid not. In April of this year, NEWSWEEK magazine reports that Putin is popular with 21 percent of Americans," I say.

"Here in Odessa, I run from political talk and bite my tongue often. But how can you like Putin and support Ukraine at the same time?"

"Don't get depressed," she says. "Keep dancing."

"There are some rabbitholes of public opinion that I'm afraid to go down," I say. "I wonder. How many Americans approve of Hitler?"

12

Lying in the Shelter

Late tonight in Ukraine, the air-alert siren blows.

The black night sky scares me worse than our daylight alerts. So I try clowning to my shelter crowd.

"PUTIN IGRAT I MUY," I say in my shaky new Russian. "My rough translation is 'Putin is playing with us.'"

"Your Russian is baby-talk," the woman desk clerk says. "Speak English, better."

"Yesterday, I'm taking a shower and hear this same air-alert," I say. "Dry myself off fast, dress up and boot down to here. Putin toys with us. How come we never know about the next one?"

"Naturally," the clerk says. Like many Ukrainians, her eyes look like rare greenish jewels in a fair calm face. "He surprises us. That's his plan."

"It amazes me," I say. "How many Ukrainians speak good English."

I try forgetting about the Russian Kalibr missiles that travel at 2,000 miles per hour.

They can roar into my neighborhood from 1500 miles away, blasting my hotel and everything within a square mile.

Try running from that.

We all wonder where the rockets will strike next. Recently, rockets blew out my ballroom dancing place windows.

"You no afraid?" a young blonde mother asks me. She holds her crying baby in a blue football jersey.

Ukraine, Whaddya Gonna Do? or Sunshine, Salo and Sirens

"Why should I be scared?" I ask, lying and digging for talk to relax everyone.

"Dude, like whatever," a thin bristly youngster snaps in English. He sports two pierced nostrils, a lower lip stud and a blue throat tattoo.

"You American?" I ask. "Another bi-lingual guy. You sound like Southern California."

"Naw, man. I just watch a lotta music videos. Rap, Hip-Hop, Garbage Punk Rock. But you gotta be scared."

"Not really," I lie. "Can I tell you why?"

The clerk translates for the others.

Once more, it's time to talk nonsense and cheer up the other shelter friends.

Some look up from their phones.

Others edge near me.

Passing around popcorn and granola, I try looking wise and experienced.

"By now, we get used to these missile alerts," I say. "They come too often. We aren't that scared anymore."

"Don't believe ya," the youngster asks, "You all fake, dude."

"Dude," I begin. "This sketchy kind of a prince character in Denmark sees a gravedigger singing up a tune while the digger does a grave. This singing shocks our prince character. He asks his homeboy next to him, 'Horatio, Wazzup? WTF?'

"Homeboy Horatio got it down. Homeboy says to the prince, 'Like, man, you do some janky routine long enough, it gets mad easy.' That's like us with these air-alerts.

We survive it all the time, learn to live with them."

* * *

Shakespeare's words:

HAMLET

Has this fellow no feeling of his business, that he sings at grave-making?

HORATIO

Custom has made it in him a property of easiness.

13

'Star Wars' Scares Me

Tonight, the siren wails and scares me again. It's time to distract myself from being afraid. So I chatter with the others in the dark bomb shelter.

"Attention!" my phone says again. My phone app shows it is Mark Hamill speaking. "Air-alert. Go to the nearest shelter."

"Listen to this actor Mark Hamill from the film 'Star Wars' on my phone," I tell the group in my weak Russian.

Nobody reacts. They are my pass-in-the-hotel hallway friends. Most look too young to know the film. *Star Wars* opened 46 years ago.

It looks like nobody wants to speak English to me.

Maybe they see me as a charity case who cannot speak Russian well. I came to Ukraine to help.

Tonight, I'm asking for help.

"I trust my phone much," I rattle on. "With no phone, I do not know these air-alerts."

My friends' shadows bend, almost kissing their phones. Everyone wants distraction.

My phone buzzes.

"Attention!" Mark Hamill says on my phone. "Don't delay. Your over-confidence is your weakness."

"Scary voice to hear at three AM," I grouse, opening a fresh popcorn bag.

"I gotta say I never saw the *Star Wars* film. Hope Mark gets cash for his voice."

Tonight's bomb shelter friends are a mother about 45, her daughter of 23 and a child of six. A pale portly Turkish guy who looks like Pavarotti palavers away in the corner.

Grusa whimpers. She is a long-term hotel guest with a sorrowed pale face, missing front teeth and lank black hair, showing bald patches. She always wears the same violet T-shirt. Printed words on the shirt say, 'The Future Is Romantic.'

Shelter lights click off and on.

A young couple hauls their sleeping baby into the shelter. The baby smells of wet diapers.

I offer popcorn. Nobody takes any. I open the bag anyway. I want them to smell it and remember carefree munching popcorn in the movies.

"You eat?" Grusa asks in her chipped English.

"Grusa, it's one in the morning," I say.

"I have to work soon. Air-alerts always make me hungry. It's probably the stress."

Grusa nods. She understands this.

"DEPRESSIA," she murmurs.

On my own, I can figure that one out.

"Grusa," I say in English. "How you talk, you big-city friend! Ukraine has the worst rate of depression in the world. 78 percent of Ukraine has a relative or close friend injured or killed in this war."

Something BOOMS! nearby. The parents tense.

Their baby wails.

Up above, something booms.

It's the wrong time for me to ask about the 'BOOM.'

Frank Hickey

The baby keeps caterwauling.
The all-clear sounds.
Somehow, we turn shy and embarrassed.
I leave the shelter last.

14

And Now for Something Completely Different

"Before you found me in that Odessa Market," Kirsti asks, "what did you for female companionship?"

"Is that what we have?" I ask. "Sometimes I wonder."

"Don't exaggerate," she says. She stresses the 'rate' sound with her honey accent. So, it sounds like 'exaggeRATE.'

"I went ballroom dancing" I say.

"I can imagine," she says. "Pathetic. Banal. Hitting the hot spots, at age 70."

"Thanks," I say. "After Kherson, I found a ballroom place here in Odessa. But they only danced BACCHATA and SALSA."

"Poor baby," she says.

"In between air-raids, I wanted foxtrot, American waltz and tango," I go on. "Hungered for genteel manners, elegant music and society chitchat untouched by war. I desired all those boring stuffy things that I and my generation helped destroy. Because we called them irrelevant."

"Run home to Mommy," Kirsti says.

"Stop using your psychology minor against me," I say. "Most women dancers looked about thirty. Always hugging each other."

"That's our national sport," she says. "After sundown."

"I tried introducing myself and being a Friend to All," I say. "Handed out my name and my book cards."

"I know your mantra by now," she says. "Heard it in English and Russian. 'YA PISAL PUBLIKATENEY DESET KUNEGII, ADIN KINO, I NYET JENSHINA. I wrote ten published books, one movie and no wife.' Why don't you have that printed on a T-shirt?"

"Not many of the pretty young dancers stay in my memory," I go on. "That was ten months ago. I've been working here in Odessa a while. But I do remember walking to the dance spot one Friday evening after work in the camouflage rag factory. Turned the corner on Kanatna Street and saw something shocking. Something that I can't forget."

15

Nightmare

"Okay," Kirsti says. "Dazzle and amaze me. What did you see on your way to dancing, trying to snare our innocent young Ukranian maidens?

"Like that very horrible Brooklyn word that you use to describe a sexual deviate?" she asks.

"A horn bug," I answer. "That's the title of my second book, *Funny Bunny Hunts the Horn Bug*."

"CAUCHEMAR," she says. That word means 'nightmare,' in both Russian and French. Kirsti utters it often. "You molest our women. Like me."

"On my way," I resume, "I smelled a horrible chemical burn smell. As a New York arson detective, I had smelled that before, in horribly burned dead tenements.

"Turning the corner, I saw huge metal beams twisted like something from an angry giant's hand," I say. "An entire block, 200 feet square, lay still smoking. Clumps of locals sat on the curbs, looking dumb at the wreckage.

"A huge tan wall sagged sideways over the wreckage," I rattle on. "It looked like ripped cardboard, showing different layers of brown and gray. As if the wind was blowing the wall down as I spoke. Bricks lay everywhere."

"'Sir, this place was they home,' a bystander tells me in crumpled English. He thumbs a dirty digit at the locals.

"'Russians shoot rockets of ships in the night's middle. Want stop us, sell grain towards Africa, Asia. Kill Ukraine.'"

"Tired firemen drink coffee and chew sweet rolls near their emergency truck," I say.

"'No home now', the bystander says. "'Putin devil.'"

16

Saturday Night in Old Ukraine

Saturday Night in Odessa
"Looking around, you see no sign of war," Kirsti says. "Everybody under 40-"

"Thanks," I say.

"-wasting time as only the young can do," she finishes.

"We both see the same things," I say. "Groups of tattooed boys carrying their beer cans because they're too young or too broke to tipple in cafes. Smoking their first cigarettes. Shouts exploding. Some sporting sunglasses at night. Bearded alcoholics swaying on the sidewalk. Why do Odessa drunks always wear dirty blue jeans?"

"Their own signature style," Kirsti says.

"Some enthusiastic street musicians," I continue, "A bird dealer charging teenagers money for pictures with his rare white birds. A babble of gurgling languages, braces of girlfriends out on the town in short skirts and eye-shadow. Belly buttons showing. You would never know about the air-raid three hours ago."

"Was anyone wounded?" Kirsti pushes.

"That's a question I never ask," I say. "It's difficult for me to know. On my phone, I find the war news in English either outdated or unreliable."

"Or both," she says. "Everyone in government lies."

"That's a bleak outlook," I mutter.

"Not in this war," she says. "Actually, it is realistic."

17

Would You Feel?

"I try crossing Odessa streets on the green," I tell Kirsti. "Drivers gun their engines and zoom near me. Don't they know that I'm here to help?

"Do they drive like this because everyone gulps down CAFE AMERICANO, black, without milk?" I ask.

"Does milk calm drivers down?" she inquires.

"Calms ME down," I answer. "Wanna get back to peaceful New York. Where the weak coffee tastes like a friendly warm juice."

"Traffic runs bad here," she agrees.

We sit in our favorite cafe in green and leafy Gorky Park. Children holler nearby.

"Jumping for my life over the flagstones," I say. "I remember that Ukraine suffers the world's worst rates of mental illness and depression today. Because of the invasion. Are these drivers acting crazy?"

She looks around the park.

"Aren't we all?" she asks. "When the war started, Ukraine pulled all the police and most soldiers out of my town of Kherson. Nobody knows why.

"Russians entered Kherson. Civilians and a few Ukranian soldiers tried to defend us with old guns and Molotov cocktails. The Russian tanks killed them quickly.

"Our people marched out to meet the tanks. Holding only Ukranian flags. Teenage Russian soldiers sat on the tanks, pointing assault rifles at us while checking their cellphones and chewing gum. My friend shouted 'Shame! Shame!' at the Russians. STITT! STITT!

"The soldiers heard this and fired their rifles. I don't know if anyone was killed. The Ukranians just kept marching, holding their flags."

"When was this?" I ask.

"March 6, last year."

"March 6," I murmur. "The battle of the Alamo."

"How would you feel if Russian tanks roll down your Fifth Avenue in New York?" she asks. "And you can do nothing but hold flags? Just stand there, waiting to be bullied, ordered or killed.

"Would you be mad?" she goes on. "Would your entire country get angry and become a bit crazy?"

Now she glares at everything in the park.

"And some of your American leaders call this 'a territorial dispute,'" she snarls. "Like Hitler invading Poland? Shame!"

18

What Would You Say in the Shelter?

I sleep.

"Attention!" my Odessa phone blurts. "Air-alert! Go to the nearest shelter."

I say a terrible word.

My watch reads 3:43 AM.

Shorts, knapsack, red rescue shirt, thick boots for walking through rubble and I slop towards the bomb shelter below.

Like always, the shelter smells of neglected wet.

It lies empty.

At this time of night, most Odessa folks remain a-bed.

Part of me feels stupid for sheltering.

Part of me feels wise, deep down in my amygdala.

Grusa limps downstairs to the shelter.

She wears her violet T-shirt reading 'the Future Is Romantic.' As always, she sports fresh makeup over her pale face and brown wig. Few teeth survive in her lipsticked mouth.

By now, I can understand parts of her Russian.

"KAK DI LA?" I ask and then switch to English.

"How are you?"

"I was a ship's cook," she mumbles in Russian.

"Speak little English. GOOD EVENING, GOOD NIGHT.'"

She does a curtsy inside our bomb shelter.

"THANK YOU VERY MUCH," she lisps.

"Where your ship was?" I ask in ruptured Russian. Sometimes, I call my own language RUPRUS. It sounds like a police radio acronym.

I try to draw her out. It feels like digging into an ESCARGOT with a tiny spoon.

Grusa shrugs. She turns back to her phone.

That tiny square of light shines in our shelter and onto her face.

It is the shelter's only light.

"MY SON," she says.

Her English words sound loud in the shelter.

Time drags past.

"Attention!" my phone says. "The air-alert is over. May the Force be with you."

I straggle to my room. Grusa stays in the shelter. I don't know why.

Days later, I ask my friend Kirsti about Grusa.

"Grusa lives forever in this hotel," Kirsti tells me. "Like you."

"I thought that she works here," I say. "Because she never leaves."

Kirsti shrugs.

"I've studied psychology," Kirsti says. "Maybe the war makes her regress to her child state."

"Psychology means nothing to me," I say. "I'm here trying to help refugees. But after Grusa and I sheltered together, she started hanging her wet underthings on the hotel balcony."

"She had never done that before. Somehow, somewhere, there is a connection in that, between our shelter night and Grusa's unmentionables. What would Herr Doctor Freud say about that?"

"Don't make a joke," Kirsti says.

"She mentioned her son," I say. "I didn't ask anything."

"No," Kirsti says. "Don't ask."

19

Ukrainian Battered Women

In Odessa, Ukraine, my boss brings me to a shelter for battered women. The women victims, of all ages, stare at me from chipped wooden desks.

My boss asks me to discuss domestic violence. She interprets.

"As a policeman in America, do you see much domestic violence?" the boss asks.

"Too much," I answer. "In one eight-hour shift, my LAPD partner and I arrested three different abusers on three different calls."

"Ukraine has such problem," she says. "The war start, we have 40% more domestic violence."

"In America, when someone abuses another, the police MUST arrest," I say. "It's the law."

"Ukraine has same law," she says. "But sometimes our police do nothing."

She interprets. Women nod.

This leads me to another question.

"How often do the police here follow the law and arrest the abuser?" I ask.

"DESET PRAT SANT," one woman answers.

She bobs her dyed red hair over black eyebrows. "I speak little English. Ten percent, police do this correct thing."

The other women nod.

Ukraine, Whaddya Gonna Do? or Sunshine, Salo and Sirens

"That's no good," I say, sounding like Mister Fix-it, a naive idealist, hoping for the best. "Try this. Tell your church minister, your lawyer, a journalist or a politician about your abuser. That brings attention. Abusers and lazy police do not want attention. I have known a lot of both."

Another woman, with heavy makeup and jewelry in the Russian style, snorts.

"No good," she says. "My husband has nightmare that I am Russian soldier. He try to kill me. Police tell my husband, 'Don't do this thing again.' Then, they go. I speak Ukrainian. Not Russian. Some time, police not speak Ukrainian. Only Russian."

"Why?" I ask.

Everyone shrugs.

"Language matters," I say. "Often, I had to speak Spanish on these abuse calls."

We finish the class in the local style, with black coffee and Ukranian sweet rolls.

Then I leave for my other job.

Among other tasks, I cut and hang brown, green and gray rags for camouflage nets.

At work that week, my colleague Anna makes me smile. She says that my Russian is improving.

Anna had to say this in Russian because she can't speak English.

The day after Anna praises me, I enter work and feel something different.

When I sit down at our work-table, Anna is crying.

The other rag-cutters, all women, hunch at our table with her.

Again, I do not know what to do. Should I leave? Or ignore her crying?

In a neutral kind of way, I start cutting rags. The other women drift outside to smoke.

Anna is a single woman, 52, with glasses and a mild temper. In two months, she never shows anger or frustration. Most of the others do. Being a refugee angers them.

I recall that Anna had danced ballet. Using my phone, I turn on a ballet performance.

Maybe the ballet will comfort her. During this war, we need some relief.

Just like I don't obsess with war news. Hearing the rockets explode nearby my pillow is enough. I do not ponder the war news.

To fight obsession, I return on YouTube to the TV shows from my childhood. These are Jim Bowie, Yancy Derringer, Peter Gunn and others.

They remind me of how I saw life as a six-year-old.

They relax me.

My ballet scheme falls flat.

Anna keeps crying. She turns her phone to Google Translate. English words show on her phone.

"MY FATHER DIED THIS MORNING," the screen reads.

Maybe he died in a rocket attack. I do not have the guts to ask anything.

I have to say something in Russian.

Patrol cops give this talk all the time.

We cops try comforting the survivor.

Because we talk with the survivor long before detectives, sergeants, media or grief counselors see them.

In my police past, I gave this comfort talk in Spanish.

Today, I give the comfort talk in Russian.

20

Kirsti, Why Don't You Make Me Real?

"So, more readers are enjoying what you write about us in Ukraine?" Kirsti asks.

"I hope so," I say. "The writer never knows. I could be the crown prince of the 'Delete' button worldwide."

"Why do you call me 'Kirsti'?" she asks. "I'm a real person."

"Completely," I answer. "Totally."

We are walking down the famous Potemkin Steps, dipping down to the Black Sea. A hot April wind blows Kirsti's dark curls.

"Long ago, I explained that to Our Readers," I say.

"Explain to me again," she says.

"First, you ordered me not to write about you," I explain. "Commanded, if memory serves. Nothing about that pineapple upside-down cake of longing, love and loss that is your private life-"

"Thank you," she snaps. "Very kind of you."

"And then there is the style called 'the New Journalism,'" I gas on. "The writer may admit at the start that he will use composite characters, give his opinions and use the techniques of fiction to make his story more interesting."

"You hope so, anyway," she says.

"Hope is what keeps us going," I say. "Especially in Ukraine. So, I put the questions about our life here between you and me. In bright, sophisticated witty dialogue."

"And so modest, you," she purrs.

"Mother Hickey taught us kids that false modesty is just that," I say. "As fake as the thirtieth of February.

"So, I think that dialogue between a man and a woman feels more interesting than just reciting dry facts. Also, readers ask me 'What is their relationship? Are they lovers? Past lovers? Just cafe pals? Will they be lovers in the future?"

"I'd like that question answered myself," she says.

21

Black Sea Beach

Lying on Odessa's Black Sea beach sand, I would never know the war still rages.

Teenagers slam into each other and holler, playing volleyball. Bikini beauties yearn for admiring looks just like the tattooed weightlifter men do.

Parents hoist children into the low waves breaking frothy white. ROMA Gypsy peddlers sell water, crying out. Loud beer drinkers clog the cafes, with rock music pounding into my ear.

An air-alert siren sounds. Some swimmers stop. They look to the shore.

Odessans know that the Russians target our port with rocket attacks. Rockets have already destroyed food packaged to feed hungry millions in Africa and Asia. Rockets do not care.

The port lies next to the beach where I swim. Army bases cluster nearby. The whole area is a bulging fat target.

Lying on the sand, I try to google data about underwater explosions. From childhood, I remember that underwater explosions are more severe.

My phone reads that an underwater explosion creates a pressure wave. This wave attacks the body, blocking blood vessels, rupturing lungs, tearing internal tissue and hemorrhaging the brain.

Saying a terrible word, I slide back into the water.

Former Russian president Medvedev said last month that if Ukraine advances, Russia 'would have to use nuclear weapons'.

Maybe Putin will decide to use nuclear weapons without warning us swimmers on the beach. I wonder, treading water. Did we Americans warn the Japanese before bombing Hiroshima and killing 140,000 civilians?

22

Rockets Hit

Rockets hit the last two nights.

They pounded Odessa at two AM.

"Russians want attack us again tonight," Daria says at work. She speaks broken English, a tallish Ukrainian woman with rare green eyes and stringy brown hair. "In the same time."

I'm waiting, at five minutes to two.

It's like having a party begin. You rush around to get things in order. Then, you steal time and stretch out for a nap.

Four minutes to go.

Last night, the Russians fired missiles from ships in the Black Sea nearby. I can walk to the beach from my pillow in an hour.

Signs on the beach forbid photos or videos.

Last year, the Russians attacked the port of Odessa, hoping to seize it.

Ukraine tries blocking saboteurs near the port.

Two minutes left.

Russia wants to terrorize Odessa by these raids.

At work, I try to smile, sing to myself and clown around. I imitate John Wayne and greet the Ukrainian grandmamas by drawling "John Wayne, AMERIKANSKI cowboy GAVARIT, 'Howdy, cowgirls.'"

From Mongolia to Intercourse, Pennsylvania, everyone snickers when you invoke John Wayne.

Ukraine suffers the world's highest depression rates. Being a clown may help.

I don't want to tell Daria that Hitler wanted to terrorize Holland before invading it.

So he bombed Rotterdam. In 45 minutes, 60,000 people died.

It's now 2:09 AM.

The first rocket THUMPS! nearby.

I feel the shock in my head on the pillow.

23

Help Us If You Can

"You give money to a charity," I tell Kirsti, my cafe friend. We sit on the beach cafe, sipping coffee. "Do you ever wonder how they use it?"

"I never give to charity," she answers.

"On my own money, I fly to a foreign country," I say. "This country Which Shall Remain Nameless. It was not Ukraine. I want to help Ukranian refugees. I check into an over-priced hostel, share a room with scruffy smelly kids who leave cold pizza plates out on the floor. I make phone calls, like 'Here I am, to save the world. When do I start?'

"One volunteer czar tells me to call back in a week. A week? I'm going to tread water in this new country and hostel for a week? Why a week? Don't you know there's a war on?"

"You always say that," Kirsti says.

"That's because it's always true. Not wanting to wait more, I called another outfit. Charity X.

"The X person at the other end says that they package food for Ukrainian war refugees. She gives me their address. Beautiful. I say that I will get there."

"Maybe," Kirsti smirks.

"No maybe. For sure. Using Google maps, I find the place. About 15 miles away from my hostel pillow, north of the city.

"This was how I operated as a kid private eye at 22. In those days, if I didn't find the witness or the subject, I didn't get paid. I learned to come at them out of the morning sun, before coffee.

"So, in this new country, I rise early, breakfast and take a local bus to the huge monstrous bus terminal. Nobody knows nothing. Nobody speaks English. I'm waiting for a number 34 bus. Nothing shows.

"Asking other passengers gets me shrugs and 'Nice to meet you, sir,' in English. Useless phrases like that.

"I go to the Information booth. Long line of other lost fools there. I wait my turn. Show the written address to the sleepy wizard behind the window. Like they say in Brooklyn, he don't wanna know. Grudgingly, he points out where I can get bus 34. I go to a new spot, wait some more and climb onto bus 34.

"We drive and drive. The buildings get smaller. I'm wondering how I'm going to get back. But no matter. This war demands sacrifice."

"Yeah, right!" Kirsti tosses her black hair.

"Then what all expatriates dread happens," I say. "Bus 34 goes to the last stop, the normal people leave and the driver looks at me funny. Starts barking in foreign talk. I slink off onto the blistering hot street. Check the nearest bus terminal yard, all these buses standing around like ancient elephants waiting to be exterminated."

"Then what happened?" Kirsti asks.

24

Don't You Know There's a War On?

"Shambling around what looked like a bus yard," I say. "I knock on a Quonset hut door and surprise two fellas at a desk. I spy a computer on their desk and wave my paper scrap with the address on it.

"By now, that paper is starting to look a bit gnarly from being folded and unfolded in this heat."

"Did you feel like you are a detective again?" Kirsti asks.

"Detectives spend their lives looking for non-existent street people in empty phone booths," I say. "So, this is old turf to me. I persuade these two gallants to check their computer for my address. They do it, direct me to another bus. I take that bus to the last stop and start waving my paper at unfortunates in the street. I'm sweltered from the heat. I stagger into a store, pay too much and drink a bottle of beautiful ice water. By now, I'm so tired I'm walking crooked. One man points me to a building. I go inside and am overjoyed to find the X charity."

"Finally," Kirsti says.

"Right. I knuckle the door. It opens. A youngish man speaks fluent English. He seats me in a waiting room. The office is packed with youngsters in office clothes.

"They try not to stare at me. A brisk young woman comes out to tell me that they don't have any work for me. It hits me. They're handling me like I'm homeless. By now, I look pretty ragged."

"I bet," Kirsti says.

"I tell her that I'm looking for volunteer work," I say. "And that one of her people welcomed me and sent me this address.

"Right away, she gets solemn. Wants to know who welcomed me. I can sense office politics about to avalanche and cover me.

"She wants to see the email. She doubts my word. 'Who sent you that email?'"

25

Sweaty Fruit

"So, like a murderer proving his alibi, I show this Charity X Executive Woman the email inviting me here," I say. "With her office address. Said Executive looks like she wants to drown the woman who sent me the email. And drown me alongside her.

"I can only imagine the message sender's reception at the office tomorrow. I'm happy to be three busses and 15 miles to the south when that happens.

"She makes an executive decision and calls me a taxi. The taxi will dump me at their food distribution center. I offer to pay for the ride. She refuses. She's probably going to torture the email sender with that taxi bill."

"Why do you cause all this trouble?" Kirsti asks.

"The war. Taxi takes me off the streets, into the country, down a muddy dirt road for twenty minutes and finally to the food center. It was like a hidden guerilla stronghold. The workers there, some Americans, treat me like I had come in a flying saucer from Mars. Alongside them, I box and package fruits and vegetables. It's hot and sweaty work. After my odyssey getting there, Our Hero was not at his best."

"How did you get home?"

"One of my co-workers, didn't speak English, got orders to dispose of me ecologically. Murder might have been a temptation. He wheeled me back to the city and dumped me kinda sorta somewhere near my hostel.

"I thanked him like a good boy, limped out and rested at the first cafe for nourishment to keep body and soul together for the trek home. Panting. Dirty. Fruity. At the food center, the peaches had been somewhat mouldy. By dead reckoning and the setting sun, I got back to home sweet hostel, showered and fell into my bunk."

"What happened afterwards?"

"I never heard from Charity X again. The next day, I started farming the Internet for another worthy cause."

26

Changing Isolation

"Do you think that I should change?" Kirsti asks.

"From what to what?" I answer.

"I don't know," she says. "Maybe someone more cheerful?"

We are dancing slowly in the Odessa shelter during an air-alert. Our shelter audience ignores us, staring into their phones.

"You can try," I say. "We all should. But it's okay to be angry at this war."

Outside, a siren blares and then stops.

"I'm angry about nobody helping us enough," she says. "Big countries like Iran and India are moving to help Russia. Poland says, no rockets for us. Why doesn't your America do more?"

"Good question," I say. "Do you know that America often hid from European trouble?"

"Some of you want to hide now," she says. "When did this feeling happen before?"

"Before both World Wars," I say. "Some Americans felt that it was not our problem. They called themselves isolationists."

Outside, something goes BOOM! I flinch. Kirsti ignores the noise.

"Yes?" she asks. "Tell me whatever you know."

"George Washington started this idea," I say, stuttering from nerves. I feel shaky and weak.

"Georgie W. told young America to watch out for entangling foreign alliances. America listened up. Until World War One. We only fought that war for 17 months."

"We silly vain Europeans suffered it for four years and five months," she says. "Some countries never recovered."

"Before World War Two, we were scraping through a Depression," I say. "Recovering took everything we had. So we ignored Mussolini, Hitler and Tojo as they murdered innocents to gain power.

"We birthed a group called the 'America First Committee.' They shouted for us to ignore Europe. They boasted 800,000 members, in a country of 133 million. You won't find this in history books. Few Americans know about it. I didn't know it myself until last year. "

Behind us, a baby cries in the dark. We keep dancing.

"They opined that Germany posed no military threat to America," I say. "Other groups like them grew fast. We felt bitter over losing 116,000 soldiers in World War I and getting nothing in return. Isolation groups were progressive, conservative and pacifist.

"Big shot fat-cat politicians turned isolationist, too. They stripped our military. The Navy stopped doing full-scale drills in 1926. That left us totally unprepared for war in 1941."

"What changed America's feelings?" she asks.

"2400 dead Americans at Pearl Harbor's sneak attack changed us," I say. "The isolation movement died while the bombs were still falling. Stopped cold. But this week, some Americans want to stop helping Ukraine."

"Maybe America should change again tonight," she says. "Like I am changing."

"Like the next Russian rocket might change both of us," I say.

She looks over at the bent cell-phoners. Her throat pulses angry.

"If Russia takes Ukraine," she says, "what do you Americans think will happen next? Do you think that Russia will stop? Why should they?"

27

How it All Began, New Mexico Cowboy Detective to Ukraine

"When did you first get this noble but crazy idea to come to Ukraine to help us?" Kirsti asks.

"Right after Russia invaded here," I said. "Russian troops massed for an amphibious landing here in Odessa. I was working as a detective for the State of New Mexico Public Defender's office. Media alerted me to this invasion."

"So, what did you do?" Kirsti asks

"It's time to play the Flashback Music," I say.

"Back to New Mexico, in March of 2022."

* * * *

March, 2022

"I can't sit at my computer, watching this Ukraine invasion," I tell my boss in his New Mexico office.

"Meaning what?" he asks.

"I gotta do something," I answer.

He shrugs, a wise Black street lawyer in a rumpled gray suit, pushing 50.

"What do you want to do?" he says.

"Got 40 plus years of interviewing witnesses, suspects and victims," I say. "In eight languages. Defended the American way of life when necessary. Scared most of the time. Graduated somehow from torturous police academies staffed by muscular tyrants. Must be some way I can use that experience to help the Ukranian army."

"You making good coin here," he says. "Especially for dirt-poor New Mexico. Think about that."

"Cash doesn't rock my world," I say. "Detective work was always my first love. Started at 22. Every job before that was guerilla warfare, me against the boss."

"You're Irish," he says. "Not Ukranian. And what, 65?"

"68," I answer. "And Irish-Jewish. I can serve there. Long as they don't ask me to run too fast. Or do pull-ups. I can work security investigations, de-brief victims or train rookie cops."

"Your age, where you gonna get another job like this one here with me?" he asks.

"You're right," I admit. "Never."

His office mirror shows me in a dark jacket, silver bolo tie against white shirt. My new black cowboy boots tap the floor. A blue wool cowboy hat hangs on my office rack.

My boss had allowed me to wear the boots, hat and bolo tie to court. Everyone else does. The judge sports boots and a huge rodeo belt buckle under his robes. Because I am working as a detective for the Public Defender's office in a small town in South New Mexico. I try to blend in with the locals.

So far, blending in as a former New York City detective is not going well. After weeks in town, I know nobody. Everybody drives pickup trucks. Nobody walks or talks to strangers.

Carlsbad, New Mexico, is the most car-dependent and dangerous town I had ever gotten stuck in.

No taxis drive here. The only bus shuts at five and closes on the weekends.

And Carlsbad festers in the top 12 percent of high-crime towns nationwide.

"If you really feel you should go, you can write me out a resignation letter," he says.

I look out the window at the rocky green hills just outside of town.

This is big scary stuff. I can feel it through my wrist nerves, in the veins there.

"You know, you can get killed over there," he mutters.

28

Breaking Out

Carlsbad, New Mexico, 2022

Gifting my office wardrobe to the friendly Bangladeshi family running the motel, I try escaping Carlsbad, New Mexico.

"Ain't no bus station here," the amiable crone at the local stop-and-rob convenience store says. "Got a jitney stops on Saturday outside by my gas pumps. Goes into Texas."

"Where?" I ask.

She shrugs and drags a broken fingernail through matted gray hair.

"Texas," she repeats.

"Out of here?" I ask.

She pauses to grasp the concept.

"Yep," she says.

"Good enough," I say. "Sell me a ticket. Please."

Saturday morning, I gather my things and trudge to the store through the snow and ice-crust. Carlsbad lies near South Texas. But for some reason, winters blow fierce and long through here.

"Hey, y'all!" a beefy red-meat Hunter Type catterwaules at my store clerk. "What chal mean, 'No more jitney tickets'?"

"Thet cowboy over chere git the last one," she wheezes.

'What cowboy over where?' I wonder.

Ukraine, Whaddya Gonna Do? or Sunshine, Salo and Sirens

I look around, forgetting that I was still wearing my woolen blue beauty cowboy hat. That hat was my Ticket to Blending In.

The Hunter Type glares at me.

New Mexico lists no gun laws. Nobody registers guns. Anyone can carry a pistol openly. Except me. By law, I have to work unarmed.

Everyone in Carlsbad oozes handguns.

They drip cartridges. I think that I might be safer outside in the wintry gale.

The Hunter steps closer to me.

"Y'awl gotta ticket?" he asks.

29

Colorful New Mexico

It seems like a good time to use a lawyer-delaying-tactic from Brooklyn felony court.

"Sir," I drawl. "Yessir. I'm glad you asked me that question."

"Says which, Bubba?" he responds.

I strain not to sound like a wisecracking New York City detective. Which I was. Please forgive me, Carlsbad, New Mexico.

Back in New York, I used to crack wise and drive my more mature partners to distraction.

Remembering my boyhood adventure stories, I decide that it is time to spin him a yarn.

"You see, sir," I begin. "My grandpappy's uncle worked as a dowser, finding water in West Texas and selling cook tinwear to widder women 'long the turnpike. Spent years hunting a whiskey tree that would bear fruit and run this old-timey barbecue down in Ce-ment City, Texas, before Dallas proper swallowed it up-"

"Ticket?" the Hunter repeats.

Something clanks outside.

A yellowish metal caterpillar of a vehicle about twenty feet long oozes out of traffic. It slumps next to the gas pumps.

A huge bearded driver with neck tattoos munches yellow corn chips from a plastic sack that looks like a garbage bag.

He keeps chewing while shifting gears. The caterpillar coughs deep in its engine, sounding diseased.

"Excuse me, sir," I say. "There's my ride."

Moving light on my feet, I edge past him.

"Bubba, I'm talking to yuh!" he yells.

"Not for long," I mutter.

Watching his hands the way the police academy had taught me, I gain the door, yank it open and run to the jitney.

Inside the jitney, unhappy passengers glare at me. The bearded driver considers my aspect, crunches some more corn chips and picks the broken bits from his beard.

Then he swings the jitney's door open.

The Hunter bobs up behind me, closing in.

Pushing my duffle bag ahead of me, I swing up into the bus.

"Next stop, Ukraine," I pant.

30

Can You Pronounce the Name Pryzemysl?

My memories distract me. I close my eyes, open them and realize that I am not in New Mexico anymore. I am actually in Odessa, Ukraine.

"In this ongoing saga of YOU, what happened when you left New Mexico for Ukraine?" Kirsti asks me.

Now we are dancing under the trees in Odessa's Gorky Park. Near the amusement rides, someone plays slow jazz music.

The frozen past seems far away. That makes me happy.

"I never went wild with love over Eastern Europe," I gripe. "During my youngster homeless years, I froze in Bulgaria. Serbia in June wasn't much better. I can still feel that cold in my spine. But Putin's invasion pulls me back here."

"I know," she says. "You would prefer that he invaded Bermuda."

"Just after Putin invaded Ukraine, I'm trudging on snow like a fool through my first day in Warsaw, Poland, still wearing my New Mexico cowboy casual clothes," I say. "Believing that my idealism could keep me warm. Because my heart was pure."

"Ha," Kirsti sneers. "We mature grownups know better."

"Warsaw train station was a outdoor experience," I say. "And lovely brisk. For seven hours, I'm shivering there. No place to buy a coat was open on Sunday. Poland is so wonderfully Catholic that stores shut down at Saturday noon."

"Next time that you try saving the world, maybe you should plan," she says.

"I planned something," I say. "Arranged to sleep in a student dorm in the Polish-Ukrainian border town of Pryzemysl.

A refugee told me to go there. And the name is not pronounced the way it's spelled.

You call it 'SHEM-Mesh.'"

"That's crazy," she says.

"Not to a Pole," I answer. "After the train, at midnight, a friendly Polish cab driver in Pryzemysl brought me to a student dormitory building. In good English, he promised to wait until someone answered the door. He dumped me on the doorstep and sped away. No waiting. Liar. My next book of memoirs, entitled 'Lying Cabbies I Have Known.' '

"I knocked on the door and waited, freezing and scared."

"Why scared?"

"Scared nobody would answer," I say. "I kept waiting."

31

Will Russia Attack Me, Me, Who Everyone Loves?

"In my first hour in downtown Pryzemysl, Poland," I say. "Shivering in the cold, waiting for someone after midnight to open the student dorm door."

Kirsti spins me around, dancing.

Ukranians gawk at us dancing in Gorky Park.

"The door opened and I fell in," I say. "Some ghostly student type whispered to follow him. He unlocked a gnarly cubbyhole and I tumbled onto the bed kind of thing. It felt warm. Don't remember anything after that.

"As soon as I could, I ankled down in the frosty sunny morning to the train station. It was a huge greenish stone building, built in classical Roman design.

"You could see the refugees," I say. "Their crumpled lined faces showed the horrors of war. Dark Roma Gypsy families staggered through the station.

"Volunteers wearing different colored vests walked among groups of refugees. The refugees lugged huge stuffed suitcases or plastic leaf bags across the floor. I saw one large pale guy, shrewd eyes and short muddy hair over a golden vest, looking calm and capable. So I asked him about his job.

"'We do translations,' he says in a soft Polish accent. 'Carry refugees' luggage, feed them and run a safe sleeping shelter.'"

"'Put me to work,' I said. And he did. For hours. I enjoyed it. But I needed a room close to the station.

"He drives me to a hotel along the Sar River. Refused payment. 'No, no,' he says. 'You're my friend.' Just the same, my paranoid detective training made me press my palm near the car's gas tank. That way, if he kidnaps or robs me, my palm print will identify the car."

"You are not normal," Kirsti pronounces.

"Want to be abnormal and safe," I say. "Not normal and dead. I notice three steel girders in a triangle about six square feet. To my eye, they looked ugly.

"'They call them 'hedgehogs,' he says. "Built to stop Russian tanks in World War II'".

"'We don't have that worry now,'" I say. "Ukraine stands between Russia and us here in Poland.'"

"He looks at me like I broke loose from somewhere. Gets a wild look in his eye. Too late, I realize that I am an unknown in a war zone and I just upset someone. And I'm in his car.

"'Don't you Americans know anything?'" he asks.

"'The eternal question,' I mutter.

"'Poland borders Russia!' he shouts. 'Up north in Kallinigrad! Nothing stops them! They could invade us tomorrow, just like they did in Ukraine!'"

32

Strong Forts

"In Pryzemysl, Poland, behind the metal anti-tank hedgehogs, a round stone house squatted all obscene below ground level," I tell Kirsti as we dance in Odessa, Ukraine.

"It was a blockhouse for Polish guerillas to live in and fight from. Gunports and firing steps. Old blood smears. Rotted wood."

"How morbid," she groans.

"Pryzemysl suffers from a morbid history," I say.

"Later, I heard that in the 1880's, the Austro-Hungarian army built dozens of forts in hilly Pryzemysl. They used the hillsides as natural defenses against enemies. Attackers would have to charge uphill into the forts' cannon fire."

"Were the attackers that dumb?" she asks.

"Patriotic," I say. "Full of 1914-style enthusiasm. Some French generals called it 'ELAN.' To some, it means 'Fighting Spirit.' Acting fearless caused so many deaths in the war's early days. 70 percent of all helmets were wooden. Remember, the civilian leaders called it 'the War To End All Wars.'"

"It didn't," she says.

"We moderns know that today," I say. "But most felt differently about their country then. Nobody predicted this would lead to more than four years of ugly and pointless trench warfare, sharing wet straw pallets with rats, corpses, snow and flooding.

Ukraine, Whaddya Gonna Do? or Sunshine, Salo and Sirens

"Here in Pryzemysl," I say, "during the first two weeks of war, the Czarist Russians attacked those forts and lost 40,000 troops. With my dicky knee, I walked those same stone forts and underground passages, full of the last breaths of dead young soldiers who died full of ELAN."

"Why would you want to work without pay in such a depressing cold town?" she asks.

"To stop it from happening again," I say.

33

Poland

"Kirsti," I say. "Once again, you're right. Prysmel did feel like a depressing town."

"What did you expect?" Kirsti asks.

"But our work still feels vital," I say. "This war awakens me like a fire-bell in the night. We were the first railroad stop for refugees fleeing Ukraine. Some came exhausted, their tongues hanging over their shoulders. My group set up a safe shelter for women and children.

"We fashioned our shelter as an outdoor plastic structure in the frosty spring nights," I rattle on.

"Sometimes our portable heater worked. Mostly not. Refugee women huddled with their babies, crying.

"But it was better than sleeping on the platform. That town always felt cold. Men and boys over 17 snored in another area, as the Polish police in black uniforms and slung assault rifles watched them. After dark, some men refugees get romantic. And like all train stations everywhere, we dealt with hustlers, thieves and drunks. The drunks always shouted the word 'KURVEH!' whenever life vexed them. It was kind of the all-purpose Polish cuss-word and they wore out my ears shouting it."

"That word means 'a loose woman,'" Kirsti says. "Language is similar to Ukrainian. Typical macho-man philosophy, to curse a woman."

"Men were always trying to sneak into our women's shelter," I say. "Some refugees had nothing. Once, I hoisted a senior refugee woman's huge plastic shopping bag from the train. The bag broke open on the platform."

"What was inside?" Kirsti asks. "French ticklers?"

"You're learning Americana way too fast," I say.

"No. The bag was packed to bursting with Ukrainian movie magazines. Nothing else. What do you think of that?"

"When you're fleeing murderous genocide, you take what you need," she says.

"That's not funny," I say.

"Depends on your philosophy," she says.

"Was Pryzemysl jolly to look at?"

"The San River looked pretty in sunlight," I say.

"Glistening cold water rushing past. Lovers and lone storytellers like me shivered on benches. And the train station always fascinated me. From what I could understand, they fought the Eastern Front's last battle of World War I, over this marble and brick-and-green tile train station. Without many cars back then, trains kept troops moving and away from murderous trench warfare. Artillery barrages. Here, in my train station, we struggled to handle Europe's worst tragedy in 80 years."

"What about Human Trafficking, sex slaves?" Kirsti asks. "Did you see any of that?"

34

Sky Pilot Fights Back

"Let me play policeman again," I tell Kirsti. "What can you tell me about Human Trafficking?"

"It happens everywhere," she says. "And women lure the women victims into it. Because women trust other women."

"Right," I say. "From hamlets in Thailand to luxury Manhattan condominiums near the United Nations. With diplomats trafficking."

In Odessa, we are dancing in our tiny frigid bomb-shelter during an air-raid. On Kirsti's phone, the Glenn Miller band plays the jazz song 'Perfidia.'

"Said Trafficker finds vulnerable women or children," I drone on. "Easy in war zones like Ukraine. Offers them a job and housing. Then the Trafficker locks them into sex slavery."

"How can the Trafficker fool educated women?" she asks.

"Simple," I say. "Trafficker is a citizen who speaks the language. The refugee has none of that. They are vul-ner-able, with a capital 'V.' The Trafficker piles them into a van and drives them to a hidden place.

"Now, the refugee feels lost. No idea where she is. Trafficker says if the refugee tries escaping, she'll be killed. Along with her kids. And the Trafficker says that they bribe the police and judges so nobody can touch them."

"Can you do anything, as a civilian volunteer?" Kirsti asks.

"Civilian chameleon," I say. "I can change colors. When I saw someone that I suspect of trafficking, I tell my group. They pass the word to other volunteer outfits. We relay information about them."

"You sound like another useless cop," she says. "Just gathering data."

"Oh, yeah?" I growl like a New Yorker. "I gathered my data to a Sky Pilot in the station."

"What the devil is a Sky Pilot?" Kirsti asks.

"A religious leader," I answer. "Holy man. That's old cowboy slang. Each time that I did this, the next day, said Sky Pilot would supply me with the person's car, address, nationality and anything else necessary."

"How?" Kirsti asks. "By the power of prayer?"

"That's what I asked him," I say. "He followed them. Just like a private eye. And handed their data to those burly Polish cops with the shiny assault rifles.

"Poland, real religious country," I continue. "Sky Pilot asks, cops do. Cops say, 'Yes, Sky Pilot! Right away!

"Then cops tramp out all happy and religious and prolly put the boots to the suspect. Dance on his head in some dark alley at night."

"You're just guessing," she says. "Tell me something. What did YOU do?"

35

Walking Against the Traffic

"The UN says that this war displaced 78 percent of our Ukrainian children," Kirsti says. "So soldiers kidnap them to Russia. Or Human Traffickers recruit them for sex slaves."

"When I worked in Poland, I saw that," I say.

"In the frosty railroad town of Pryzemsyl, we set up a safe sleeping shelter for refugees. Kept them away from Human Traffickers who were always looking for new slaves."

"How could you tell if someone was a Human Trafficker?" Kirsti asks.

Kirsti and I are still dancing alone in our bomb-shelter to the Frank Sinatra song 'All my Tomorrows.' This air-raid is going into extra innings.

"There are ways that you can tell," I say. "Especially if you are a genius detective like me."

"And so humble," she says.

"One subject type of guy kept coming around my train station," I say.

"What's a 'subject'?" Kirsti asks.

"The word 'subject' is jargon from my former days," I answer. "We gracefully aging characters use jargon from our glory days of youth."

"So, if he is your subject, that makes you king," she says, smirking.

"Call him whatever you want," I say. "Call him 'mutt,' 'crook,' 'Human Trafficker' or 'slimeball.' I approach to engage him in learned speaks."

"Whether he wants it or not?" she asks. "Is that legal?"

"Sure," I say. "I'm a dumb friendly American volunteer. Gregarious. Whaddo I know, Polish criminal code? Somehow, all Traffickers speak scraps of English."

"I see Traffickers here in Ukraine, too," she says.

"Listen to our talk," I say. "You tell me if it's legal. Now:

> Me: Hiya, I'm Franky, American volunteer. I noticed you here at the station for a few days.
>
> Can I help you with anything?
>
> Subject: No English.
>
> Me: Aw, I think ya do. Heard ya talk before.
>
> Subject: Excuse me-
>
> Me: What's your name, sir?
>
> Subject: Don't understand -
>
> Me: So, I see ya got no luggage. Never get on a train. Only talk to women and children. Why is that, sir?
>
> Subject: Don't have, talk with you. Because you are only volunteer.
>
> Me: Sure. See this picture in my hand? That's me. Los Angeles Police Department. Inspection day. Don't I look sharp? Very aggressive, tough police. They fired me. Said I was crazy. Do I seem crazy to you?
>
> Subject: I go.

Me: Of course. Sure. See ya tomorrow. I've enjoyed our chat. Catch ya later.

Subject: I tell police of you.

Me: Somehow, I doubt that. Do you want that much attention from the cops?

•.   *.   *.

"Did the police really fire you?" she asks.

"No," I say. "But they did call me unusual."

36

Singing Opera During Air-raids

"Kirsti," I say. "Guess what happened today?"

"Tell me," she says.

"I'm alone in my bed," I say.

"I should hope so," she answers.

"'Attention!'" my phone blares near my bed.

"'Air-alert! Go to the nearest shelter.'"

"I say a terrible word. My watch reads 2:21. In the morning.

"Do you think that writing about this will interest the reader?" I ask Kirsti.

"Not this reader," she says.

"Throwing on the Polish army raincoat, a light parka, pants and slippers, I shoulder my emergency backpack," I go on. "Bandages, popcorn, tools and flashlights rattle together.

"Down in the bomb-shelter, the same group of young musicians from China are sitting. Last week, we had talked Mandarin. Tonight, they bend over their cellphones, like yearlings drinking from a stream. Looking scared."

"What do you expect?" Kirsti asks. "A Dayglow light show?"

"Outside, something goes BAHHHRR!" I say.

"I flinch. They flinch, too.

"'You need a distraction,'" I say in English to this Chinese group."

"'Take your mind off stuff. Ya need a rodeo clown.'"

"Puzzled looks flower.

"Taking a deep breath, I launch into 'CHE GELIDA MANINA' the aria from Puccini's opera 'La Boheme.'

"Maybe they know it, maybe they don't. This is improvisational music theater.

"A large Chinese man joins my aria.

"We sing it together. He sings one line, I follow him with another.

"He looks about 23, wearing a raggedy black shirt showing holes. Sings the aria perfectly. His crowd relaxes. Some clap."

"I can't believe this," Kirsti says.

"I posted our aria on Facebook," I say. "Do you want to see it?"

37

Teaching What? What?

"It's starting, Kirsti," I say.

"It certainly is," she says. "You're going more mad?"

"Impossible," I say. "I'm trying to teach my students. They keep asking me 'What? What?' What's starting is that I'm thinking of myself as a teacher."

"Not as a policeman?" she asks.

"No," I say. "Not some flatfoot hair-bag dragging that piece of lumber across Life's Great Stage."

We are leaving the Odessa Opera House, after seeing the ballet 'Don Quixote.' Kirsti wears a lollipop-red evening gown, her styled black curls and golden sandals. I cannot catch up in my local thrift-shop tuxedo and dress shirt, black cargo pants and worn patrolman shoes.

"I assigned my class the Walt Disney film about Elefgo Baca in New Mexico, 1884," I say.

"Wanted to show anti-Latino racism, then and now. Do you know that today 25 percent of Latino American college students experience discrimination? That same percentage consider dropping out of college to avoid more discrimination."

"Really?" she asks.

"Next class," I say. "One student says, 'I watched it. This show is just a typical Western.'"

"And he's an expert?" she asks. "In Odessa?"

"Kirsti, right then, I suffer a detective flashback,"

I go on.

"Someone is lying to me. Trying to tell me the story of Goliath and the lion. Or Who-Shot-Willie nonsense. Who hit Annie in the fanny with a flounder? Damn student didn't watch the film."

"So that hurt your new professional teacher pride?" she asks.

"Somehow, yes," I answer. "So I ask the student,

'Tell me about this typical Western filmed in 1958. Where the hero is a Mexican gunfighter, outwits corrupt judges, beats White racism to become a sheriff, lawyer, mayor and district attorney. And it's all true history.'"

"You can't pressure students like that," she says.

"Really?" I ask. "I remember getting about 14 years of such Catholic pressure in my schools."

"So cute that you care," she smiles.

"Does my class understand that Disney was the only person to ever win 26 Academy Awards?" I ask.

"He shaped American culture. By blood, he was English, German and Irish. Not Latino.

"But he was an animation genius who used his power to create America's first mass-market Latino hero in Elefgo Baca," I say. "And he did this when most White Americans saw Latinos as bandidos, tango gigolos or migrant fruit-pickers."

38

Odessa Da-Da Sky Pilots

On Sunday afternoon, I limp across Odessa's City Garden Square, when the air-raid sounds.

The Square forms a green tree park with an orchestra gazebo and fountain.

It always feels like Odessa's center, surrounded by overpriced cafes and stringy kids shouting over their folk guitars.

A trio of Religious Missionaries start folding up their stand and heading out of the park.

This piques my curiosity.

Always the witty conversationalist, I spring into action.

"You're these religious types, right?" I ask in clear English. "Missonaries. Sky pilots?"

I can see them figure out my words.

"DA, DA," the man says.

He is wiry and dressed well in a gray suit that looks hot in the heat.

These words always tickle me when Odessans answer 'DA, DA,' for 'Yes, Yes.'

If memory serves, Dada art came out of anger at World War One. Dada called itself 'anti-art' and used nonsense in paintings, music and sculpture.

"I respect your religion," I say in my baby-talk Russian. "All religions. Your philosophy."

They keep packing.

Rapidly.

"Why do you leave the park?" I ask.

"TREBOHA," the man answered.

That word means 'air-raid' in Russian.

"But nobody in Odessa does much with air-raids," I rattle on. "Only schoolteachers with their children in the park. They run to shelters. Everyone else forgets them."

The trio begins forgetting me.

They leave the park.

My wise guy Brooklyn detective genes start taking control and launch into a soliloquy.

Just like Shakespeare.

"Why do you characters slide outa the park during air-raids?" I ask, practicing my Russian. "The danger is outside the park.

"Buildings will fall on you. No buildings in the park here. What's the story? Don't you sky pilots believe in your own product?"

39

"You Think You Know Your Neighborhood."

"You think that you know your neighborhood," I tell Kirsti. "And then Putin cuts the juice. The lights go out."

"What's your new mumbling?" she asks. "Crazy?"

We pick our way down a sloped flagstone street.

Odessa just cut the electricity.

Neither of us can see the flagstone sidewalk beneath our shoes.

"You don't know your area until the lights go out," I say. "Then ya gotta feel your way home. Like now."

Russia keeps rocketing Ukrainian power plants into ashes.

So Ukraine cuts power to save up for their harsh winter.

We struggle on.

The air-raid siren blows.

My left foot hits something hard.

"O showers of bastards!" I intone. "My ankle!"

"Your mouth," Kirsti says.

A car headlights bleach white the street and then move on.

Odessa's French and Italian-style buildings stand over us.

Dark tones of pink, gold, jade and butter glow.

We struggle for safe footing. I smell something familiar.

Feeling my way, I remember that Ukraine has a murder rate of 4.3 for every hundred thousand people.

To compare, America has about 6.2 murders for the same group.

Ukraine, Whaddya Gonna Do? or Sunshine, Salo and Sirens

Ukraine is not Chicago.

But it ain't Switzerland, either.

"We don't go down that street," I tell Kirsti.

"Why not?" she purrs. "I must get home. Work, work, work!"

"I grasp the philosopy," I say. "However, smell that street's marijuana fragrance?"

"So what?" she asks. "I'm not smoking it! Not anymore."

"But we don't know who is," I say. "And we can't see him, courtesy of Mr. Putin. And Friend Smoker mayhap is a scoundrel. A raspscallion. He grow discourteous if molested.

"Let us promenade this away in tranquility and feel our way home. We'll live longer. Even during this air-raid."

40

Policing Meets Young Ukraine

"You were a policeman in America?" my boss in Ukraine asks me. "Can you teach young policemen?"

"Hope so," I answer. "Me, a teenage preppy-hippie turned street cop. Been trying to teach younger cops for years. To keep us all alive and happy in our job."

Two days later, I'm in a police university, in front of 200 students, boys and girls, ranging from 16 to 20 years old.

Some will graduate to become detectives.

The others will be prosecutors.

From what I know so far, Ukrainians do not rate their police highly.

They often call their police 'corrupt'.

Changing this is a challenge.

It might start in this room.

"In 1990, my city of New York suffered 2245 homicides," I speechify.

A woman professor translates alongside me into Ukrainian.

"Six murders per day," I say.

"Everyone saw this as inevitable. Life in the big city. We all suffered and learned to live with it. Then a new police commissioner, Bill Bratton, changed their techniques."

"What did they change?" a willowy woman student with tinted blue glasses asks.

"They used the crime-prevention theory of 'Broken Windows,'" I say. "Picture a city home with broken windows. Nobody cares about it. The city never fixes it.

"Years pass. The tenants leave. Kids paint graffiti under the window. Vandals steal pipes. The homeless break in to live there. Small-time marijuana dealers sell nearby. Nobody stops them. Armed crack dealers take over. They will shoot each other and kill innocents nearby. This leads to gang wars.

"All because of one broken window. I'm simplifying, of course-"

"You certainly are," an unseen student murmurs in English.

"-But that's one theory," I continue. Nothing daunted, that's me.

"It seems impossible to believe that," the student answers.

"The police attacked crime by fixing that window," I say. "Sure they used other techniques. Can we talk about them? But the murder rate dropped fast. Last year, we had 386 murders. Not 2245."

41

'Well, my Mind is Going Through Some Changes.' (Buddy Miles, musician, 1971.)

I'm talking with cadets in the Ukranian university for police detectives and government prosecutors.

A professor at my elbow translates.

"In American policing, what would you change?" a woman cadet in a gray Minnie Mouse T-shirt asks me.

"This is my own idea," I say. "Every month, everyone in the department sits down in private with a psychologist. From the chief on down to a temporary summer clerk.

"They speak for just five minutes," I rattle on.

"If everyone has to do it, there is no stigma attached to it. No gossip. Because most professionals understand how much stress policing causes. Just listen. You will hear it.

"If any problems arise in this short talk, the psychologist schedules another session, a longer one, as soon as possible."

"Does any police department do this?" she asks.

"Not that I know of," I say.

"Why not?" she pushes.

"I've asked police chiefs and supervisors that same question," I answer. "In my view, that would cut down on complaints about brutality, excessive force, shootings and racist acts.

"The police bosses usually agree with me," I say. "This psychological talk would prevent problems. But when the department finds a troubled employee, what can they do with him or her? Desk work? There aren't enough desk jobs in most departments."

"Why not?" she asks.

"Because 75 percent of American departments have less than 24 officers," I say.

"Desk jobs go to officers who are physically injured, obviously ill or too old for the street. There just isn't enough desk work to put psychologically troubled cops there as well."

The class of 200 cadets stirs.

Some look up from their cellphones.

"Police bosses tell me this," I say. "'If we find an officer with actual psychological problems, we will have to fire him. And he might sue us for his pension and win. A jury might agree that we caused the problems. We would lose thousands of dollars. And what about officers faking stress to get a pension?

"'So, the best thing that we can do is screen them psychologically when we hire them. Watch them in the academy and during probation. After that, we just have to hope and pray that they don't develop severe problems and kill any Innocents on camera. That's what terrifies most of us police supervisors.'"

42

"Bombings Morph to Backups," I Say

"GOVNO," she says in Russian. It is a bad word. "Why do you say that?"

"Because in the Los Angeles cops, I would be in the station, processing an arrestee," I explain.

"Some cop's voice would shout on our belt radios. '415 man with a gun! King and Crenshaw! Need backup!'

"Everyone would forget paperwork, jump up and run for their cars. Boots pounded. Cusswords sounded. Blue wool rasped. Leather gun belts creaked. We all raced to get there first."

"If this is a metaphor, I don't see it," she says.

"Look at my room here," I say without thinking.

"Must I?" she asks.

"You're right," I answer. "It IS a bit artistic."

Rain pelts my hotel room window. Kirsti looks at plastic water bottles lined against the wall.

Tiny dwarf notebooks show new Russian verbs that I had scribbled. Scrappy Ukrainian socks dot the floor.

"But when the air-raid sounds, I don't have to think," I say. "That is the metaphor. It's my emergency routine. "

"Kirsti, these air-raids are like…"

"You talk in parables?" she asks. "Think that you are Jesus?"

"Pull on the long pants with passport, cash and cellphone in the patch pockets," I say.

"Slip on shoes. Shrug into the heavy coat because the shelter always blows cold. Air-raids form a fast dance."

"Why fast?" she asks.

"Because Ukraine cannot defend us anymore," I say. "They don't have enough missiles. More Russian rockets get through their defenses. How much time do we have to reach the shelter?"

"I don't know," she says.

"Nobody does," I answer. "I don't want the building falling onto my head and me dying here, not knowing the answer to that question."

43

DREAMS

"Kirsti, you know where I'm waking up these days?" I ask.

"Not near me," she says.

"In the bomb-shelter," I say.

We are food-shopping in Odessa's famous outdoor PRIVOZ market. Heaps of mysterious porky meats show white marbling alongside golden cheese mountains. Tattooed soldiers buy rain ponchos and daggers at the surplus stores.

"You should be in the shelter," she says. "If a rocket hits your hotel, you can't cha-cha dance your way afterwards to a long life in America."

"Oh, you hear that I'm squeezing dance lessons into my English classes?" I ask.

"Naturally," she says. "In Odessa, everyone gossips. Often, about you. Crazy American."

"Anywho," I say. "I wake up in the shelter and check my cellphone. There was no alert all night. Did I dream this alert and come to the shelter for no reason?"

Kirsti snorts.

"Getting older," she says. "It may be dementia. You should check your phone before going to the shelter."

"That's no help," I answer. "Because I always do check it. I do that both in real life and in my dreams. Sleeping alone, how do I know which is real and which is a dream?"

44

During Air-raids, Playing 'the Blue Danube.'

"Why do your phone always play that Blue Danube waltz during air-raids?" Kirsti asks.

"To remind us of civilization," I say.

"No," she says. "Why?"

"You ask why?" I say. "This is why."

In our air-raid shelter, I seize her arm and spin her into waltz position. Cultured Viennese style, we waltz.

Creamy notes flower and rise.

The hotel guests pass us by. Some smile.

But others show nothing.

After two years and four months of bombings, drone attacks and mines killing beach swimmers, nothing much surprises Odessans.

Maybe the guests are afraid that we're going to pass the hat for cash assistance. So they keep a stone face. Smiling might get expensive.

"In Mexico, they sing this song when they want cake at birthday parties," I say. "'QUEREMOS PASTEL, PASTEL!'"

"You're goofy," she says, trying out her Yankee slang. "You know more things that don't make you any money than anyone I know."

"This Blue Danube waltz is also a shrewd humanitarian carol," I say. "After Russian rockets destroyed millions of grain in our Odessa port during July, the merchants decided to sail their grain ships down the blue Danube instead of onto the Black Sea."

"You believe that Ukrainian propaganda?" she asks.

"Ukrainian grain feeds millions of natives in Africa and Asia," I say.

"To make money," she smiles.

"Most people, in Ukraine or somewhere else, crawl outa bed on rainy gray mornings to make money," I answer. "Others, some geniuses, write what seems important here in Ukraine. And give their writing away. For free."

"You're a genius?" she purrs. "And so modest. Is your writing worth anything?"

45

Wedding Chit-Chat

"Three air-raids between midnight last and dawn today," I say to Kirsti.

"I hope you went to the shelter," she says.

"Hope is what keeps us alive," I say. "Around two AM, I curled up on the shelter sofa and slept until the all-clear. About three hours later."

"Senior like you," she says. "Must be all tired out today."

"You betcha," I say. "But this is war. You sleep when you can."

For once, we are dining in a good restaurant.

Generous windows show us the Odessa streets.

Tattooed teenagers sluice past, sunlight glinting on their nose rings.

Her cellphone buzzes.

"Boyfriend?" I suggest in a neutral kind of way.

"Oh, it is that terrible man from Baluchistan," she says. "He calls to marry me. But he lets the phone ring thirty times. Why?"

"Animal lust," I say. "The only possible explanation."

"I visited my grandmother in Dubno," she says. "You texted me. You were sick with flu. I wanted to help heal you but I could not."

"Take a walk after dinner?" I suggest.

"I can't," she says. "Maybe I will sleep early tonight, Monday."

"Everyone says that on Monday evening," I say. "Hedge-fund millionaires and homeless folks sleeping on cardboard. So, no walk."

"Oh, I can't," she wails.

Her cellphone buzzes again.

"O," she wails. "It's him again. Thirty-times-calling-man. What can I do?"

"Tell him you're saving yourself for Hollywood," I say.

"You think that I seek to distract you with him?" she asks.

"Take a walk?" I ask again.

Salesmen say that repetition works sometimes.

"No," she says. "How many times must you ask this question?"

"You're great when I'm sick," I say. "When I'm healthy, you always find other things to do. But injured or sick, you're a faithful friend. If Putin blows my leg off, you'll probably want to get married."

46

Bread Line, Water Line, Riding Circuit

Today Odessa feels more desperate. Heat wave wraps around the city. We suffer electricity and water cuts. The air-raids keep pounding us.

Tonight, I walk a gritty part of downtown Odessa with scrappy parks. Shouting panhandlers scrabble through gray dumpsters.

A water line of about 300 locals in sweaty summer clothes form up at the public water station. They grip boxy bluish plastic bottles.

Nobody drinks tap water here. The Russians have smashed $7.5 billion of water purification systems.

A thin tattooed man is haranguing the crowd with angry Ukrainian words. To my ear, he seems to be ranting about the war.

In America, Mother Hickey told me about bread lines during America's Depression of the 1930s. Starving men crumbled and fell on line.

Then she watched her own father's face shrivel and age. My friend Kirsti shakes me out of memory.

"Why do you think that you working here helps us?" Kirsti snaps.

"I'm riding circuit," I answer. "Like judges and preachers did in country America in 1890. Crude knowledge and a desire to help. Just doing what I can."

"Do you notice that your favorite cafeteria uses only paper plates now?" she asks.

"No," I say.

"Because they don't have the water to wash their porcelain plates," she says. "I worry. What's going to happen to us this winter?"

47

Ukrainian War Warp Speed-Dating

"Kirsti," I say. "'The New York Times' just wrote a piece on Ukrainian dating during this war."

"Did you learn anything, Mister International Police Association Life Member?" she asks. "Detective work about our women. Remember, I gave you permission to date other women."

"I remember," I answer.

"You must learn about Ukrainian culture," she says.

"I'm learning," I say.

We are dancing to slow jazz in Gorki Park. Odessa women smile at us. Younger Odessa strides by, wearing tattoos and hunched over their phones.

"What have you learned so far?" she asks, simpering innocently. "I bet that some young girl cheated you."

Her smile grows, like a Cheshire Cat.

"I learned that when you take a lady doctor out for a first date," I say, "she orders two drinks, you order coffee and the bill comes to 1200 hrynia. Or $40 dollars, US."

"Whatever did she order?" Kirsti asks, keeping her eye on the competition.

"How would I know?" I say. "She whispered her order to the waiter in fast Ukrainian. That must have been some drink."

"It sounds like the doctor operated on you," she says.

"That week, I taught my English class the meaning of the words 'Clip joint,'" I say. "It's a term from old Brooklyn. A clip joint is a place where they sell you 100 hrynia of something and they bill you for 500 hrynia.

"When I saw the bill, I made sure that we danced to the live piano player, after the painful operation of paying," I say. "I insisted that we dance. Maybe it was tacked onto the bill. I could hear my mother's voice from decades ago, saying, 'Frank, don't be cheap.'"

"Maybe her mother told her the same thing," Kirsti says.

48

On the Road Again

"Back in New York for two weeks and I don't have to sleep with my wristwatch on," I tell Kirsti.

"No?" she asks.

We are fumbling my rolling suitcase near the Odessa bus station. Different hustlers scan us. Their knife scars and wolf eyes remind us that Odessa suffers crime.

"No wristwatch," I say. "I don't have to know what time it is when we lose our lights and electricity. It doesn't happen in New York."

"Only in Odessa!" she snaps.

"No air-raids," I gas on. "Speaking truly, I don't have to sleep with anything on. Completely naked."

"NESTI," she murmurs. That's Ukrainian slang for 'You're talking trash.'

She's right, as usual.

"No emergency wardrobe lying on my bed in Ukraine," I say. "No backpack filled with water, popcorn, crackers, charger and flashlights."

"Aren't you tired of chomping popcorn during air-raids?" she asks. "I know that I am."

"Some people got vexed at me for leaving Ukraine," I say. "Others snap that I shouldn't go back to Ukraine. 'This Ukrainian war will end when Trump gets elected,' they say.

'Why risk your sweet self over there for those people? You're not even Ukrainian.'"

"They sound like intelligent Americans," she murmurs. "Do they still exist?"

"No side issues, please," I say. "Gaze upon this bus station. Is everyone genteel and scholarly?"

"If you don't like it here, why come back?" she pushes.

"That's a powerful line of inquiry," I answer, jockeying my bag past a Panhandler Who Has an Unhappy Look. "In New York, sometimes I didn't know why I was there. And today, I don't know why I'm here."

49

Horse Sense

"ATTENTION!" the actor Mark Hamill from *Star Wars* hollers on my phone. He wakes me after midnight. "Air-alert! Proceed immediately to the nearest shelter."

Grumbling, I roll over in my bed.

"Mark, ya need an editor," I growl in Brooklynese.

"Foreigners don't understand that word 'proceed.' How about 'Run directly to the nearest shelter?'"

Emergency shelter clothes fold onto me.

Windbreaker, long pants, strong shoes for stepping through rocket rubble and the backpack with water, nuts, popcorn, charger and books.

We live with dead neighbors and shattered homes.

My window lies open in this heat wave.

Far below, I hear something unusual.

CLIP-CLOP. CLIP-CLOP.

It makes a slow sound.

Nothing else sounds like horse hooves striking stone. Back in America, I grew up hearing it.

Scared but exhilarated, I rode horses three times a week for three years. Saddle leather and sweaty horse flesh smells stayed with me.

In Odessa tonight, a walker guides a big bay horse along the street. They walk near my window. The horse SNUFFLES! through his bridle.

Ukraine, Whaddya Gonna Do? or Sunshine, Salo and Sirens

"Horses don't care about air-raids," I mutter, stumbling towards the shelter. "Or conquering new territory. Or that election overseas.

"And how to pay the Netflix bill. Horse just wants a loose cinch, cool water and good oats to eat. Can we learn from horses?"

50

Fighting Women's Run Fu

"The Russian soldiers use sexual attack as a weapon of war," I tell my English class of women in Gorky Park. "And they could invade here tomorrow. Most often, a soldier is taller than you. So, we attack from below."

The women stir nervously.

"Stomp the foot," I say. "Nobody expects a foot stomp. How can the rapist defend his foot? Or the shin. The knee. Kick the knee or below. Then, what's our favorite Asian fighting art?"

"RUN FU!" they chorus.

"Right!" I preach. "RUN FU! So run! Most karate or jiu-jitsu schools teach useless and unrealistic moves. They are a business. Have to make money or shut down. So they demand that you pay an initiation fee, buy their costume, shell out for a locker deposit and purchase the teacher's book.

"Runaway Capitalism! Nobody in our class today is making any money. Especially me."

They giggle at that. They always do.

"We teach what works," I gas on. "No attacker can protect his feet. We surprise him.

"Many police outfits train cops in useless fighting moves so that the criminal clients don't sue them," I speechify. "I know. Been in too many academies."

"How many?" one student asks.

"Nine," I say. "But nobody's suing anyone here. The Ukrainian courts are too busy."

The women, from 14 to an age where I'm afraid to ask, practice foot stomping.

"The soldier who attacks you may have friends, weapons, experience and drugs in his blood," I say. "Could be psychotic.

"So, we don't want a long fight. We hit and run."

"Why you teach this?" a woman economics professor with solemn eyes asks.

"Fighting always fascinated me," I answer. "Even in my prep-school dormitory. Maybe because I always lose fights. Even now."

They make understanding noises.

The professor smirks.

"We have an expression in English that we should learn," I say. 'Child is father to the man.'"

"With Frank," the professor says. "Child is father to the child."

51

Odessa August Routine Days

Routine and I were always enemies.

When I ran away from home at 17, I wanted every dawn to birth an unpredictable, jeweled, rare, 'Ulysses,'or 'Romeo and Juliet' day, lasting until tomorrow's sunrise. That's how we feel at 17.

Today, in rag-tag, air-raid Odessa, my work schedule firms up into a routine. Stringing cloth in the camouflage rag-factory for the army or peeling potatoes for the homeless, I need other challenges to keep going.

So, I am teaching seven days a week. Struggling for clever, I mix my English classes with book studies, ballroom dancing and womens' Jiu-Jitsu.

Caving in to my own taste, our book club reads Walt Disney, Raymond Chandler, James Baldwin and others. We discuss the themes and the plots.

One of my childrens' classes is joyful, riotous and loud. The other class is too quiet for my taste and that makes me wonder why.

I never know or ask who has suffered trauma from this war. In my view, all of us in Ukraine have suffered.

Me, least of all. Because I can always leave the country. My students cannot.

52

Ukraine August Songs Cheering Our Hero Up

"I just came back from Greece," my Odessa friend Quentin tells me at our beach in his Dutch accent.

"Everyone was more relaxed there. More cheerful."

"Why not?" I ask like a Brooklyn wiseguy. "Who is trying to bomb them? Nobody."

Quentin agrees with me, looking out over a crowded beach afternoon. His blond hair and gold glasses catch sun rays.

Hundreds of towels protect bikini bodies under modest umbrellas.

"The war in Ukraine can color your days dark," I tell him. "Traveling somewhere would help."

"But you can't travel with your teaching schedule," he reminds me.

"Most of the happiness that I see seems to involve fast loud cars or drinking," I say. "Maybe this strikes me over-clearly as a non-drinking pedestrian.

"And the stress turns me tense, too. Crossing streets, I have to dodge reckless drivers. I could start shouting at them. Or at the idiots zooming past me with motorbikes on the sidewalk. I have to thread a path to calming myself. Sometimes singing works."

"And just what do you sing?" he asks, in his professional doctor's voice.

"To fight the war blues, I try singing to myself," I answer. "My signature song is 'Over There.' Other Americans sang that, fighting for Europe in World War One. I spread it everywhere.

"And songs like 'This Could Be the Start of Something Big,' the 1958 broadcast with Steve Allen. 'Moon River,' 'Wait For Me, Baby,' by the New Vaudeville Band. 'Please Come to Boston,'"

"Does anybody alive know these songs?" he asks. "Aren't they quite obscure and forgotten?"

"I'll give you obscure and forgotten," I answer. "I'm singing back to my childhood. I sing the second verse of the TV show 'Have Gun Will Travel.' How many of you readers out there know that there is a second verse?"

53

Fear Too Much

"'Everybody Loves a Clown,'" I tell my Ukrainian class. "Please Google that song."

They do.

"As a policeman, I was no Superman," I say. "I often felt like a clown. Once I was chasing a thief. I jumped out of my police car to catch him. A car fender hit my leg. Who do you think was driving that car?"

"Don't know," my young student Clara says.

"Nobody," I say. "It was my own car. I had not used the brake the right way and jumped out too soon. So, I hit myself. But with all my clowning, I learned some things."

"What did you learn?" Clara asks.

"For one thing, when a Russian soldier attacks you, is he fast or slow?" I ask.

"Why do you have this fear always?" Lianna asks in precise English. "And contaminate us with it?"

"You just said it," I answer. "I'm afraid for you."

"Why?"

"Because, as a clumsy naive policeman, I worked with sexual attack victims," I answer. "Most feel shame. They blame themselves for wearing that dress, going down that street, dating that man or not learning how to defend their bodies."

We breathe in the park air of fresh-cut grass in our Saturday morning class.

"I was no great cop," I say. "And I hope that you never get attacked. But that shame can lead to guilt, depression and suicide, even years later."

"I always carry tear gas spray," she says. "In my purse."

"Get it out," I say. "Fast. One, two, three-"

She unslings her purse.

"-four, five, six-" I drone on.

Fingers paw in her purse.

"-seven, eight, nine," I push.

She whips out a tube.

"-ten, eleven, twelve," I finish. "That's twelve seconds. The man has already attacked you. It's late, you're on a dark street and we ain't here."

"You taught us not to say 'ain't,'" another young woman giggles. "But it's cool hip-hop talk. Like 'yo, bro!'"

"Nevertheless," I laugh.

"So, what do you teach us?" Lianna asks.

"Stomp the foot," I say. "Kick the shin. Or knee. Then, what's our ancient Asian art?"

"RUN FU!" they chorus.

"This teaching us to defend ourselves," Lianna probes. "This is your obsession, yes?"

"Probably, yes," I say. "We police see too many victims."

"That makes you not normal," she says.

"Absolutely," I say. "Some of us retired cops drink too much, talk too much, marry too much or fear too much."

54

Critics Boo My Notes or
What Would Mother Hickey Say?

"Why do I keep sending out these pieces on Odessa?" I ask Kirsti.

"To kill my reputation?" she inquires.

"Don't be grandiose," I say. "Remember that you're not even a real person. You're a composite. A creation of the New Journalism."

"Thank you passionately," she says. "So you profit by me, marketing yourself as a brilliant international writer about this war."

We are steaming on the beach through a warm afternoon. On the stone jetty jutting into the Black Sea, bronzed fishermen cast their hopeful lines.

"Dunno how brilliant I am," I say. "Some readers reject my Odessa sketches. And they tell me about their decision."

"Very polite," she says. "Why could they not just delete your compositions, if they wish to?"

"Maybe they're trying to improve me as a writer," I say. "Through punishment. Remember, I never know who reads my stuff. I got readers everywhere between Dahlanzhad, Mongolia and Barstow, California. It's possible that, worldwide, I reign as the crown prince of the 'delete' button."

"So, why do you keep writing?" she asks.

Ukraine, Whaddya Gonna Do? or Sunshine, Salo and Sirens

"Remember, Mother Hickey was a Broadway actress," I say.

"Ah!" she pouts. "Is it time for another Broadway tale of unknown genius and shattered dreams?"

"She trained all of us kids," I say. "That the show must go on. Just because half the audience boos you, throws tomatoes and walks out, you keep on acting your routine."

Kirsti tosses her dark curly head, looking out to sea.

"The droll thing about it is that someone critical in June praises a different piece in August," I say. "Why?"

"Maybe it depends on their mood in August," she says. "You are like their little pet doggy. But there must be a deeper, more Freudian reason why you keep writing. What is it?"

"Your Ukrainian army rejected me," I say. "They say that I am too old to fight the Russians. So, I use my words. This is how I fight."

55

Chance Conversations

"In Odessa, sometimes I see the same person three times a day," I tell Kirsti. "That makes me wonder about your Security Service of Ukraine. What about the chance conversations with strangers on the park benches or the beach? Maybe they were security agents, watching me."

"Probably," she says. "You know our history?"

We are munching on something that Odessa calls ' a 'hot dog.' It is a toast roll holding a pink sausage, usually undercooked, slathered with mayonnaise and ketchup. Carry me back to Nathan's at Coney Island. Please.

I remember that just 33 years ago this week, the Russian spy agency, the KGB, pulled out of Odessa.

Ukraine had won its freedom.

Last week, Ukraine expelled the Russian Orthodox Church from its country.

The Russian church supports the troops invading Ukraine and says that waging this war "washes away all their sins." Ukrainians believe that Russian Intelligence uses the church as a cover.

"Secret police scare me," I tell Kirsti. "As a cop, I suffered some few partners who hungered to work like that, brutal and freewheeling.

"They wanted no rules, civilian oversight or video cameras," I rattle on. "None of these jokers had ever served in any military. Maybe they thought policing was like a video game. Usually they wound up fired, jailed or both."

"That's a good beginning," she says.

"But Odessa is not Los Angeles," I say. "If some uniformed weightlifter bully takes my phone and squeezes me into a car, things will look grim for Our Hero."

"So, how do you avoid any brutal or corrupt police attention?" she asks.

"If they exist," I protest.

"They exist," Kirsti insists.

Kirsti could find corruption in church bingo.

"I avoid police attention by going against my New York upbringing," I say. "I don't litter and I don't jaywalk. And I keep praying all regular."

56

Ukrainian Cuisine

"Daddy Hickey always gave us four kids huge breakfasts," I tell Kirsti over our breakfast at my hotel cafe. "Jones Company sausages, fried eggs, pancakes and doughnuts. What do you call breakfast here in Odessa?"

"Coffee," she answers.

"And nothing else," I say. "And when your afternoon meal finally arrives, it is often bland food. Do you know how I deal with dull cuisine?"

"No idea," she answers.

From my pocket, I extract a clove of raw garlic.

"UGH!" she wails. "Garlic. Your poor students in English, cha-cha dancing and womens' Jiu-Jitsu."

"May I lecture?" I inquire politely.

"Could I stop you?" she asks.

"Medical science says that, all things being equal," I lecture cheerfully, "if you skip breakfast, you have a 33 percent higher chance of suffering a heart attack."

She scans my plate of chicken, ham, vegetables, toast and, of course, mustard. Happily, it looks just like a New York breakfast.

"You're in no danger," she announces.

"That's why I like this hotel," I say. "Where else can I eat hot vegetables at seven in the morning?"

"With your wife," she says.

"I'll make the jokes, if you please," I say. "There never was a wife."

"And probably, never will be," she says. "You think you are Captain America."

We scan hefty hairy men in tank tops and tattoos lining up at the coffee machine.

"Please regard those gentlemen," I say. "And you will appreciate me more."

"Good luck," she says.

"Remember that I came here to help you win the war," I say. "And I cannot do that on coffee for breakfast.

"In 1770, a Scottish sailor escaped prison and fled to America," I say. "He wanted to help our war against England. He became John Paul Jones, father of the American navy, our country's best sea captain. When the bigger British ship was destroying his schooner and demanding his surrender, he answered 'I have not yet begun to fight.' That became our most famous war motto."

"And how did he feed his sailors?" she asks.

"Big breakfasts," I say. "He let them FRESS slowly."

"What is this word 'FRESS'?" she asks.

"It's Yiddish," I say. "The language of some Jewish people. It's means 'to eat a lot.' It's not a genteel term. Comes from New York slang."

"Was John Paul Jones Jewish?" she asks.

"No," I say. "But part of my family was."

"And garlic prevents heart attacks?" she asks.

"Ask your doctor," I say. "He'll agree. Yesterday we suffered eight air-raids.

"Rockets destroyed an electric plant," I rattle on.

"We lost power all day. Exhausted, I walked up ten stairways to collapse onto my hotel bed. So, I need these big fatty calories to help Ukraine."

A hairy tattooed waiter tries to seize my plate.

"UZZ DAZZ," I say in my baby-talk Russian. Then I try English.

"Sir, please wait," I say. "I have not yet begun to FRESS."

57

Suspicion Torments my Heart

Years ago, I walked police foot patrol in South-Central Los Angeles. As an aging naive White ex-preppy, I learned a bit.

With other things, I learned not to write in my notebook. For some reason, that jarred the locals.

"I know that you're not writing me a ticket!" was the usual contribution to a greater society.

This confused me. Then I discovered that Mother Hickey was right. She had called Los Angeles 'a car culture.' Everyone worried themselves sick about their cars. Rich or poor, White car owner or Third Worlder, nothing else seemed to matter. Car, car, car!

In Odessa today, a rocket attack hits a power station. We lose electricity all day. To save my cellphone power, I write in my pocket notebook.

Odessans stop and stare at my notebook. But half of Odessa bends over their cellphones all the time. Pushing twin baby carriages, darting on foot through carnivorous traffic or balancing an open laptop, they tap-dance through the day on their phones.

"Why does my writing in a notebook get some people nervous here?" I ask my bearded friend Bogdan. He walks his blue-eyed Siberian Husky dog near my school.

"Because we don't know who to trust," he says, smoking a pipe with sweet-smelling tobacco. "Oleg Kulinich was the

Ukraine, Whaddya Gonna Do? or Sunshine, Salo and Sirens

head of our Ukrainian Security Services in Crimea. For years, he was working secretly for Russian intelligence. Other traitors in the same bureau were working with him. They planned to overthrow our government after the invasion."

Again, I feel stupid.

"I try avoiding political talk," I say.

"Until 33 years ago, Russia owned us," he says.

"With their power politics and secret police. Do you know what was our most popular dog back then? Dachshunds. Do you know why?"

"Because Dachshunds don't eat much?" I guess.

"No," he says. "Because nobody notices Dachshunds. They are anonymous. Bigger dogs draw more attention. With the government watching everyone all the time, nobody wanted to stand out by having a big dog. So, they bought Dachshunds."

58

Surrender? Like Holland and France?

Waves of the Black Sea make 'waffle' sounds as they slow-kiss the shore.

Last week was National Ukrainian Flag Day and Sunday formed Ukrainian Independence Day, from 1991. Flags, street corner speeches and beer parties.

Today, Odessa catches her breath.

Three Russian air raids this morning hit an electrical plant. Charcoal-gray smoke uncurls on the shoreline. We lose power. Generators grunt pig noise outside cafes to keep the coffee steaming and the beer frosty.

On the boardwalk, someone plays a recorder at high pitch. It takes a minute to place the tune of "Michael, Row the Boat Ashore."

Men limp past on the boardwalk. Neat little holes show through the crewcut hair. To my eye, the holes look like 7.62 millimeter, the Russian army round. In Los Angeles, we cops scooped up the same assault rifles every week.

Between the blue jean sky and the frothy whitecaps, we should remember that this war drafts grandfathers. Ukrainians marry young. The average Ukrainian soldier is 43 years old.

By contrast, the average American serviceman is 26 years old, from private to general.

"I don't know if anyone died in today's attack," my friend Valeri says, shrugging his dark tattooed shoulders. He swims and beachcombs every day, with his civilian metal detector. "But my neighbors talk about us surrendering. Then the war may stop."

"And it may not," I say. "I won't talk politics. But I might blab about history. Both France and the Netherlands surrendered to the Nazis in 1940.

"Four years later, Nazis were still executing French and Dutch schoolchildren whenever partisans attacked any German anywhere," I say. "That same year, Germans sent Anne Frank to a concentration camp. She died there, a month before the British liberated the camp."

The recorder keeps assaulting my eardrums. If I could locate said performer, I might request a different anthem.

We settle on a beach bench. Below us, teenage girls in bikinis slam their hands and forearms in a volleyball game. Their bodies gleam with suntan oil.

"More than 100 of you Americans have died fighting in our army," Valeri says.

"I tried to join up," I answer. "Even though I'm about as military as Lady Gaga. The recruiters laughed, called me 'DEDUSHKA!' GRANDPAPPY!"

"I've lost friends in this war," he says.

"I read that 28 million Ukrainians have lost friends or relatives in the past 900 days of war," I say, starting to preach. I try slowing down.

The recorder notes keep climbing, high and squeaky.

I wonder who is this tone-deaf religious fanatic.

"But if Ukraine surrenders, what will stop the Russians from doing more atrocities?" I ask. "What will the Russians do with all these girl-children on the beach here?"

59

Polly-Wolly Polly-Tics

"Here in Ukraine, we pay attention to your American election," my student Karolina says, in the park class. "Because the results could get us all killed. What can you teach us?"

"That's easy," I say. "Avoid politics in class."

"Didn't you say that we could ask you anything?" she asks. She sparkles as some Ukrainian women can.

"Did I say that?" I ask. "I must have been young and wild and full of whimsy."

"So, teach us," she says. "Some of us, you're the first American that we ever met."

The class nods. They range from seven-year-olds to grandmothers.

"If you insist," I say. "I never could understand politics. But I always like to read. And I discovered some things reading about this election."

"For example?" she asks.

"The American states with the worst health services are voting against improving them," I say.

"The average person in Mississippi dies at age 71. Roughly, that is also true for Alabama, Louisiana, Kentucky and West Virginia. They promise to vote for the right to die young."

"Are those states Democrat or Republican?" she asks.

"Google it," I say. "I'm trying to teach you young characters research methods."

They Google it. Some giggle.

"Where do you Americans live longest?" she asks.

"Massachusetts, Hawaii, Washington State, Minnesota and California," I say. "I checked. Those folks die at age 79, on the average. At my age, that matters."

"I won't ask the same question about those states," Karolina says. "I know the answer already."

"I worked more than six years in the Deep South," I say. "Private detective and policeman. But there are some things this preppy Yankee just will never understand."

"What else?" she asks.

"39 percent of American women have four-year college degrees," I gas on. "Men, only 36 percent. Why is that odd?"

"Because women traditionally stay home and mind children," she murmurs. "While men work. But why not now?"

"I got smart students here," I say.

She beams.

"Maybe because recently, some young American men shout that college is a waste of time," I say.

"That it's academic thought control. So, they work instead. At low-paying jobs. But women know that college gives their children a better life, health care and safer places to live."

"Personally," she asks, "do you think that college is a waste of time?"

"Personally?" I come back. "I don't wanna drag my colorful torn resume into our class here."

"Please?" another student, Mona, asks. Her blue eyeglasses glint against reddish hair.

"Okay, personally," I say. "College saved my life. I scored both my BA and MA at age 30. Before that, I was outa school for nine Christmasses and driving fast on a job-market superhighway to nowhere."

They nod.

"And, if you wanna worry about America's future," I say. "70 percent of men without high school diplomas are unmarried. Don't most of them have children?"

"Naturally," Karolina smiles. "Passionately."

"What happens when that romance turns sour?" I ask.

"The men run away to another woman," Mona says. "My grandmother tells me this. And she is always correct."

"Some fathers may stay," I say. "The honorable ones who care about their children. But is 70 percent a big number? If the no-diploma father leaves, what's going to happen to those children?"

60

What Would You Buy for Free?

Is anyone following me here in Odessa?

46 years ago, ignorant, innocent and unarmed, I followed a subject for nine days through America's murder capital, New Orleans.

With no partner, cellphone or plan, following this man exhausted me. Each night, I put him to bed in his hotel. Waking up early, I had to gamble when he would emerge for the joyous new day.

Today, in Odessa, I still remember foot surveillance rules:

> Always stay behind your subject.
>
> Never make eye contact.
>
> If you lose him, return to his home. Sooner or later, so will he.

Does the possible someone, anonymous government nobody, following me know these rules?

"Who pays you?" new language students always ask me.

"Nobody," I answer truthfully. "NITKO. I charge nothing for my lessons.

But do they believe me?

Centuries of Russian terror kill trust. Their liquid eyes say that somebody is paying this gently aging goofy American. Somehow, somewhere, somehow. Nobody works for free.

But all I do is teach English, ballroom dancing, Womens' Jiu-Jitsu and now French.

And, oh, I forgot. I write these little sketches that you are reading now.

Maybe that's why someone is following me.

61

Dvorak's New World Symphony

"When you characters land in London, New York or Canada," I ask my Saturday morning class, "what work are you gonna do?"

"I shall labor in I.T.," Rana, the woman says. "Information Technology." She fingers her tongue stud with a fingernail painted gold.

"With your Starbucks English?" Stanislaus asks her in his stilted accent. "On a tourist visa? When you have no family or contacts to help you? Maybe I shall be a bike messenger."

Around us in the park, workers in gray uniforms cut the grass. We can smell the sweetness. A rocket cut our electricity again. Cafe generators grunt pig noises.

"Bike messenger in big-city traffic without medical insurance?" I ask, trying to sound objective. "Do we see a problem there?"

"Maybe a restaurant job," he concedes.

"At least you'll eat regular," I say. "And who will you be working with?"

"If they are like me, they have no work papers," Rana says. "No documents."

"But we are war refugees," Stanislaus says.

"The Internet says that America has accepted only 170,000 Ukrainian refugees since the war began," Rana says.

"Out of 39 million Ukrainian population. So, we may have to work illegally."

"Who will you be working alongside?" I ask again.

Nobody says anything.

"My Russian is terrible, baby-talk Russian," I say. "If I stop volunteering and look for paying work here in Odessa, what will I be able to do?"

"Dishwasher," Rana smiles, "with no so grand a vocabulary. Or you may clean toilets. Like you say, this becomes steady work."

"And I cannot understand British accents," Stanislaus says. "Learning that will take my time."

"I don't like working around strange people," Rana says. "Makes me nervous. That is why I do I.T."

"Toronto, Canada, is one of the world's most diversified cities," I say. "All races live there. The same is true for New York and London.

"You mean, that we shall labor with the people of Mexican?" Stanislaus asks.

"And Egypt," I say. "Nicaragua. Haiti. Senegal. They sure won't be from Odessa. And they might resent your white skin. For many historical and traditional reasons. You'll have to learn their culture and rules damn quick."

"And what if we cannot learn their behavior rules quickly?" Stanislaus asks. "If we offend them?"

"Like my police sergeant used to say, 'Stand by to stand by,'" I say.

"Meaning what?"

"Punishment," I say. "And how might they punish you?"

"Isn't this depressing?" Rana asks.

"They won't send you a nasty email as punishment," I say. "Or insults on Instagram."

"Perhaps they will strike us," Stanislaus says.

"And kitchens are dangerous localities. Knives, water very hot, metal tools-"

"I am correct," Rana says. "You men are depressing."

"I don't wish to be melancholy," I say. "But I should be teaching you more than English irregular verbs. I want you all prepared to survive in your new world."

62

Sunset Dancing on the Black Sea

"When you dance alone on the beach here in Odessa, what do you think happens?" I ask Kirsti.

"They transport you to the crazy house," she answers.

"Not yet," I say. "When I did it tonight, nobody paid much attention to me."

"Ah, my poor dear," she says. "Doesn't your show business family DNA require an audience?"

"Wise woman," I agree.

We watch the final daylight torch our beach with grayish tones. Party animals noise up the boardwalk.

Cigar smell mixes with painful rap music.

"I did forty-five minutes non-stop of my jazz-ballet-Zumba mixture," I say. "Nobody stirred. They probably thought that I was whacked out on some drug or mental disease that science hasn't dared to name this far."

"Logical," she murmurs.

"When I danced on beaches all across Asia for two years, do you know what happened?" I asked.

"I'm afraid to ask," she says.

"Groups of young MACHO men drinking beer would approach me," I say. "And in terrible English, they would ask me, 'Why you dance alone? You are gay?'

"'Not yet,' I would answer."

"Brilliant response," she says. "Did women ever ask you that same question?"

"Oh, yes," I say. "But more genteel. They would ask just 'Why you dance alone?' And I would ask them right back, 'Would you like to dance with me?'

"'Oh, no!'"

"'That's why,' I would answer."

"I'm sure that they're still discussing you in Mongolia," she supplies.

"But tonight in Odessa, nobody made eye contact," I gripe. "Like Mother Hickey used to say, 'My audience was sitting on their hands.'"

"So why do you dance like this?" she asks.

"Why did I come to Ukraine?" I ask. "Same reason. To help out. To make everyone forget the war and giggle a bit more. When they can. When I dance tomorrow, what do you think will happen?"

"Tomorrow?" she asks. "Transport. To the crazy house."

63

Ambassadors to Normal

"Everyone felt a shock when Russia invaded Ukraine," I tell Kirsti. "Even I felt the shock, 10,000 kilometers away in cowboy, gun-happy New Mexico. It numbed me as I tried to go through my detective work day."

"Mister American detective, the outside world is busy forgetting about us," Kirsti says. "Inflation, Israel's war on Gaza, your November election. By definition, 'news' is something new. Ukraine is s not new news anymore."

Kirsti surprises me by talking about the war. Usually, she hides her feelings about it. Emotion thickens her accent. It makes her English choppy. We are eating pistachio ice-cream cones at Odessa's Arcadia beach. Green trees stand solemn over us. Below us, the Black Sea gleams.

"Try to act normal proximate my crazy family," she says. "My work people. Annoying neighbors. Or else, I wake up screaming in my sleep."

"How do you manage that acting normal?" I ask.

"I try pretending that today is just another dull morning," she says. "That I sleep good. Eat a boring breakfast. Don't want talk about air-raids, my nightmares if my children are safe. I make a fake smile, like the Ukrainian ambassador in Washington. I fight every day to act normal."

64

"Something Medical Hit Me on Saturday Night."

"Kirsti, something medical hit me on Saturday night," I tell her. "Felt very tired."

"Old age?" she asks.

"That was my first thought," I admit. "But it's been three days now and I still feel aged and cranky."

I'm lying on my hotel bed, trying to recover. Kirsti frowns at me from the room's armchair. I'm trying to sleep. Tired feelings come and go.

"Cranky Franky," she says. "Could it be Covid?"

"Got tested today," I answer. "The PCR. Whatever that is. Could be 'People's Chinese Republic' for all of me.

"Nobody at the Covid shop spoke my New York English talk. So I hadda do my Renglish charades and sign language."

"Ah, yes," she says. "What you call your 'Renglish.' 'Russian English.' Or your 'Ruprus.'"

"Yeah," I mumble. "'Ruptured Russian.'"

"Do you think that your readers in Mongolia or Massachusetts want to read about you feeling ill?" she asks.

"Faithful readers, yes," I say. "Because when I feel ill, I get all literary."

"How?" she asks.

"I ponder Marcel Proust, a bed-ridden invalid longing for dawn in his book 'Swann's Way,'" I say. "Lord Byron, dying of fever during the Greek War of Independence. Robert Browning writing to his love, 'Be near me when my light is low.' Jim Bowie, dying of fever during the Alamo."

"This man Bowie," she asks, "he was literary?"

"Suffered from WELTSCHMERZ," I say. "You know, 'world pain.' Why the writer Johann Goethe had his character young Werther shoot himself over unrequited love."

"Are you suffering from this WELTSCHMERZ?" she asks.

"We all are," I say. "Everyone in Ukraine is. To some degree. World's highest rates of depression."

"How did you deal with this before?" she asks.

"As a street cop, when you feel bad," I say, "you'd try talking to another cop about it. He'd look at you funny. You realize that you had talked too much. Uh-oh! You got trouble."

"Why?" she asks.

"Because maybe he's gonna blow on you to the sergeant," I say. "Old Sarge won't understand. He's no holistic 'Kumbaya' therapist.

"Sarge will plant you at a desk where he can watch you for any irresponsible actions. Or have you directing traffic at midnight in an alley. It's better not to talk when you feel blue."

"But you're telling your readers," she says.

"That's okay," I say. "I'm retired. Old Sarge can't put me in that alley."

65

Yesterday Odessa's Humidity Hit 85 Percent.

So, thunder burps woke me at night.
But they were not thunder burps.
Strings of white pearl lights floated across the dark velvet sky. They exploded. Pockets of heat exploded.
Russian drones were attacking Odessa. I saw then from my bedroom window.
CUFF! CUFF!
It sounded like a gentleman coughing polite into his handkerchief.
They scared me. Brave, I ain't. As a cop, I worked gunbelt-to-gunbelt alongside brave women and men and I know the difference betwixt them and me.
These polite coughs could blow me without warning right into Aging Idealist Limbo.
Depending on your belief system.

66

Odessa Traffic

"These Odessa 'walk' signs run too fast," I say.

"For who?" Kirsti asks. Her Ukrainian accent thickens as she breathes cognac on me. Behind her, gray-white waves crash on the tan beach.

"For me," I answer. "I'm used to the 'walk' signs back in New York. They give you 18 seconds to dash across Lexington Avenue near Grand Central Station.

"The Odessa signs read '18,' too," I say. "But it feels like 10 or 12."

"Awww!" she whoops. "Don't ask me! I haven't been this silly drunk since University."

"You're letting off stress from last night's air-raid?" I suggest.

She bobs her curls, smiling.

"Good idea," I gas on. "But you're the intellectual here. Why does time pass quicker as we age?"

"Lemme Google it," she says, craning her head over the phone.

She sways.

"Here we go," she says, reading from her phone.

"This worthy, Dr. Cindy Lustig, says that as we age, we process information faster. And that makes our sense of time feel different. It moves faster."

"I'm glad to hear that something speeds up as we age," I say. "Feels to me like most of my equipment is slowing down."

"Cause you're so much older than I am," she says.

"Thanks," I answer. "But you're an imaginary composite creature. Allowed to live only through the approved techniques of the New Journalism. Therefore, you can't wound my feelings. So, my times seems faster, huh?"

"Means you'll die sooner," she giggles. "Does that frighten you?"

"Doesn't it scare everybody?" I ask.

"Some more than others," she slurs. "And you?"

"Sure," I said. "Hell. The place the holy men, the mumbo-jumbo men, the sky pilots, warned us about."

Looking over her at the Black Sea waves reminds me of death. The deep unknown. Writers like Percy Bysshe Shelley drowned this way.

The waves' power reminds me of nature that we cannot stop. Like death.

"We studied this phobia in psychology," she says. "It's called 'Thanatos.' Overwhelming fear of dying. Why are we afraid of our existence after death? We aren't scared of our existence before being born. We know nothing about it. The whole thing was painless. Aren't these two conditions the same thing?"

"You're asking me?" I say. "Better you find yourself another intellectual."

"I wrote a paper on Existential Death Anxiety," she slurs on. "We humans, the only living creatures who know that they're going to die. And humans have designed one single basic mechanism to deal with it."

"Religion?" I ask.

"No, silly," she smirks. "Denial. We deny that it will happen. And to seal this idea, some of us break sex rules, violate boundaries, have manic celebrations, harm others and try amassing wealth and power.

"Do we know many people like that?" she asks.

67

Talking through the Goat Rodeo

"Thank God, that sniper in July did not kill Trump," I say. "Government by gunfire never works. Murder is always wrong. How could the sniper get so close?"

"Because he was a smart man?" Jan, 17, asks in his British accent.

In the park near us, gardeners cut grass and haul it away in noisy carts, making RACKETY! sounds. My English language students stir.

"I got a different idea," I say. "Since I'm an ex-cop bursting with his own philosophy, the police communication that day was a goat rodeo."

"Excuse?" Jan asks.

"A mess," I say. "A soup sandwich. Politicians are investigating what happened. It's a 'he said-she said,' word salad.

"Eleven different police agencies," I gas on, "and nobody ever wrote out a plan. Nobody did enough talking. Do we civilians want police to communicate well?"

"Naturally," Daria, another student, puts forth. As usual, she smiles like a sunburst. "They must."

"A more truthful answer might be: that depends," I say. "In my rookie cop years, the veterans told us to have big ears and a small mouth. Police culture stresses that cops should shut up and obey orders. Nobody encourages anyone to speak up. And those rookie rules stay in place."

"Aren't you going to teach us some English today?" Jan asks.

"Eventually," I say. "But something else. Clear talking. Communication. That's why I try teaching you English.

"Some of you, boys or girls, may be fighting sadistic Russians next year in ice, snow or dark blizzards," I go on. "With wet socks and the flu. Talking clearly in Ukrainian might save your lives more than shooting your guns."

"How?" Jan asks.

"Because for years, I've seen good seasoned cops mis-communicate in gun calls," I say. "Watch some YouTube videos on your cellphones. Someone will shout out 'Didja search that suspect?'"

"And the cop will ignore the question," I say. "Or mumble so that nobody can hear him."

"You always tell us not to mumble," Daria says, smiling. "Say that we're too shy. Or passive-aggressive."

"Because the cop did not search the suspect," I say. "He forgot to. He was in a hurry. Or he thought that somebody else had searched him. Maybe he's focused on something else."

"And what happens?" Daria asks.

"Somebody gets shot," I say. "Bang-bang."

68

Exploding Head Syndrome

Sometimes, in Ukraine, I start to sleep and I hear a loud noise.

I wake up fast and shaky.

What happened?

"This worries me," I tell Kirsti, walking along the Black Sea beach. "It feels like I hear a loud BANG! Is it my imagination?"

"Perhaps," she says in her liquid Ukrainian accent.

"But sometimes, right after, the air-raid sirens blow," I say. "So, I heard the BANG! for real. My phone blurts out the 'Star Wars' actor Mark Hamill's message again: AIR-ALERT! PROCEED TO THE NEAREST-"

"Please," she says. "I know. We all get the same message."

"And, other times, I hear the sound, and there is no siren," I gripe. "No air-raid."

"So, you imagine this," Kirsti says in her psychology student tone of voice.

"Could you please check the Internet about this hearing imaginary loud noises ?" I ask. "As a foreigner, I get no medical insurance here. Companies refuse me. Civilian senior in a war-zone? Goodbye! Don't let the door hit you where the Good Lord split you."

Kirsti bends her dark curls over the phone and reads.

"This disease is called Exploding Head Syndrome," she says.

"Huh?" I respond.

"'Reticular formation in the brainstem,'" she quotes.

"Aw, I don't like the sound of that," I gripe. "At all. Those sound like cold words. Wintry."

"Do you perhaps prefer the term 'parasomnia,'"? she inquires, reading aloud. "Or 'episodic cranial sensory shocks'"?

"Worse," I say.

"Well, Frankushka, you're an educated senior law enforcement professional-

"Staring into a bleak old age," I add.

"-so what do you really want?" she asks.

"I want my Brooklyn Mommy to tuck me into bed," I say, "and give me warm Ovaltine to drink so I can sleep."

69

Looking Good, Mark Hamill

"I am in my work yesterday," Daria says, her dark brows knitting in worry. "I hear BUMP! BUMP! We look outside and see two white spots in the sky.

"Our defend stop drones," she says. "Russian drones want to kill us. Because we are Ukraine."

My morning class twitches, hearing her. Palms hold cellphones.

"Good English, Daria," I say.

"But bad Russians," Misha, our class clown, giggles. His blond hair sways over innocent blue eyes. The others smirk. "Gen-o-cide, I think!"

"'Genocide' is the same word in English as in Russian," I say. Trying to speak slowly and clearly is hard for me, a wisecracking fast-talking New Yorker. But a teacher should set an example. I think back to Mr. McLaughlin in 1963. Now I know why teachers drink.

"Frank, today," Misha says. "Do you still go to shelter for every alert?"

"Honestly, no," I say. "For some reason, I sleep through them now. Maybe I'm getting used to them."

They nod.

"When they are over, I wake up," I say. "I hear the actor Mark Hamill from 'Star Wars' on my phone saying 'The air-alert is over. May the force be with you.'"

"Yes!" Misha snickers again. "We see him, too. Class wants to know something. Why you look like him?"

"Never saw 'Star Wars'" I say. "But he does look like me. We're both Irish-Americans and about the same age. And his career is doing a lot better than mine."

"Do you HAVE a career anymore?" Misha jabs, smiling.

His audience laughs.

"Mark's raised millions to help Ukraine," I say. "Donates his voice free for air-alerts. Maybe we look alike because we believe in the same thing. Ukraine."

70

Big Apple Prep

"What I keep trying to do is prepare you characters for the New York sidewalk," I say.

We huddle under flimsy shelter of trees. Gray rain hammers the park. Overhead, charcoal clouds wring themselves dry.

"Even in the busiest parts of New York, everyone makes room," I drone on. "Millionaires edge their hips past the homeless. Everyone seems to understand that New Yorkers need room."

"The same as here?" Edward my student says.

At 12, his voice still pipes between child and adult.

"Not that I see," I answer. "Maybe you know some area in Odessa where walkers make room for others. After 15 months here, I haven't seen it. Odessans rise up outa bed and chart their return course. If you get in the way, too bad for you. They block or bump me every day."

My English class looks up from their cellphones.

"Young lovelies descend from my hotel's top-floor," I gas on. "Clerks tell me that floor is closed. I don't wanna know what the lovelies are doing up there. Because I 'm retired."

"From love?" Edward asks.

"From policing," I say. "But when I'm in the elevator and trying to exit, they push themselves inside first.

"And I give them thirty seconds of New York fast-talk in a city accent, like 'Hey, I'm tryna get OUTA de elevator before da doo-ahs close. Ya know whadda mean?' They always giggle and step aside. In New York, that scene might get a bit noisy."

"Do you worry about us?" Edward asks, his voice competing with the rain's noise.

"Yes, my children," I answer. "I admit to being a paranoid old detective when I point you towards New York. But what kind of jobs will you have there?"

"I think, low pay," he says.

"Right," I say. "And working around others like you. Low-paid Americans do not like being crowded. It offends their pride. And maybe physical pride is all they have got left. Some have done short jail time. In jail, when someone crowds you, tell them 'Gimme five feet!' Means 'Stay back.' If they still keep coming, you gotta fight. That's the code."

"Is it so violent a city?" Darka, another student inquires, a frown lining her face.

"Sometimes," I say. "The subway police trained me. 'If anyone invades your space, they are coming to hurt you. Knock them back fast.' I don't want any of you getting knifed because of a culture clash."

71

If You Love Somebody....

"If you love somebody, you must to be jealous with him," my student Daria says, in choppy English.

"You surprise me with that," I say. "Why do you believe this?"

"Why?" I ask.

We sit in my Ukrainian morning park class. Birds twitter nearby. Two unshaven park guards in black uniforms slump past, smoking cigars and beseeching their cellphones.

Daria shrugs.

"That is the natural," she says.

"Do you think that marriage counselors would agree with you?" I ask.

She smirks.

"Marriage counselors must to be all themselves insane," she announces.

Her tone vexes me.

"That may well be," I answer. "But when your marriage is turning toxic, who else can you ask for help? Your mother? Your local bartender?"

"You be ironic towards me?" she asks.

"No," I answer. "Seriously. As a policeman handling 8,000 domestic violence calls, I saw jealousy destroy many families. Love cannot live happily with jealousy."

"This only your police idea," she says.

"Okay," I say. "Then let me play teacher here. I researched this. We got 237 countries in the world.

"Before this war, Ukraine suffered the world's third highest divorce rate," I gas on. "Just behind Russia and Belarus. In 2023, after a year of war, divorces rose by 35 percent. Why?"

My students, ages nine to where I'm afraid to ask, look around. I feel stupid for asking about this.

Maybe divorce splits their families.

Later, walking with Kirsti on the beach, I pose this same question.

"You Americans do not ever understand us!" she spits out. "In Ukraine, we are traditional. That is the philosophy of the Soviet Union, still, 31 years later. Ukrainian lovers do not live together.

"Instead of living together, they marry," she says. "That is why we have high divorce rates. Because we believe in marriage. Our tradition!"

"America has the same dividing line," I say. "Our most modern city has the lowest marriage rates."

"Where is that?" she asks.

"New York City," I say. "And, for the same reason, the most traditional area of America has the highest divorce rates."

"What area?" she snaps.

"Oklahoma," I answer.

72

A Real Odessa Guy

"Kirsti, I'm turning into a real Odessa guy," I say. "At night, I'm sleeping through the air-alerts."

"You have to," she says. "Like the rest of us. We can't hurry to a shelter five times a day, like you used to do."

"But sleeping through them worries me," I say. "Is my subconscious tuning them out?"

"Heavy," she says. "Freudian."

Wind rips through Odessa today. We bend our bodies against it and slip into a cafe in the park.

"I only wake up when I hear the all-clear message," I gas on. "Then, I go back to sleep."

"Sensible man," she says, very European. "After all, what else can you do?"

"The all-clear message is a kind of lullaby," I say. "Rocks me to sleep."

"Romantic man," she purrs, like the kitten that she resembles. "And where are we going tonight?"

"Someplace romantic," I answer. "Socar gas station at L'yustdorfs' ka road, 11."

"You're taking me to a gas station?" she asks. "How romantic. You don't even own a car. Or a bicycle. Or a screwdriver."

"Socar gas stations are the only places in all of Ukraine that sell Nathan's Famous Hot Dogs," I say. "Nathan's sells 20 billion hot dogs a year, ever since 1916 in my hometown of Brooklyn.

"Your own Ukrainian hot dogs are pinkish pork obscenities, slathered in catsup and mayonnaise," I proselytize. "Entombed in some dead crusty roll. Tastes like an instant mistake.

"Nathan's hot dogs are made with a secret spice recipe handed down from the Polish owner's wife's grandmama,' I say. "With sharp mustard and sauerkraut, the hot dog makes every American tummy proud and patriotic."

"So, why are we traveling nine kilometers by bus to a gas station?" she asks.

"Bite into a fresh Nathan's hot dog," I say. "You'll see why."

73

Why Is?

Why is my Daddy's face wet? He puts his head in his hands. This is church. Does he feel bad here? He is a good Daddy. Why does he look sad?

Heat. The church is hot. We all feel it.

My Daddy, too. His freckled red face drips sweat.

We leave church.

"We need to swim," my Daddy says. "We men must stick together."

He shouts this.

My family is girls.

Except my Daddy and me.

We go to the beach.

Our car is hot.

My two aunts live on the beach. They have nobody. Their house smells doggy. My aunts never go to the beach. Rooms in the house smell closed up.

My aunts have a girl dog, Bitsy. When Bitsy dies, we see that Bitsy was a boy dog.

We go swim at the beach. My Daddy likes swimming. And the water. In the war, my Daddy was a Navy man.

Sand burns my feet. I go into the water.

I float on waves a long time. I want to be a Navy man. The beach is our world.

Girls in funny bathing suits run on the beach.

"Jeeper's Creepers," my Daddy says. That is the nice way to say Jesus Christ. My Daddy sees me looking at the girls. He laughs.

"Give it time," my Daddy says.

The sun burns my skin. It scrapes my skin.

My face feels burned. My eyelids feel sunburned. They hurt closing.

That night, lying in my aunts' spare bed, I feel the ocean hitting the beach. My eyes close. I feel I am still floating on water. I think I want to live forever on the beach in summer.

•. *. *. *. *.

Now I set my achievement goals on the Black Sea Beach in Ukraine, working a variety of humanitarian tasks during this Russia war of genocide.

My father would tell me to apply myself here but to be careful. He was never careful.

Waves hit the sand like they did in America. Far offshore, noises sound. BUP! BUP! It could be rockets or mines. Nobody knows. Swimmers stare out at the noise

Sometimes I worry myself into chest pains about a Russian rocket killing me.

Perhaps me dying from a heart attack or something else drastic and foul.

On the beach, I see wounded Ukrainian soldiers on canes and crutches.

They pull themselves along this way.

"We men must stick together."

74

As a Kid, I Wanted to Be European

They were wise, educated and graceful.

Speaking six different languages, New York Europeans would gather in cafes to discussion last night's opera. I was too shy to join their talk but I listened in.

The native New Yorkers around me seemed different. In my eye, they were foul, loud and primitive.

Now Europeans wanna act, look and sound like Americans. They wear America gangster clothes, cluster at McDonald's restaurants and stuff rap music into their cafes.

At the same time, some Odessans seem ignorant of what America is.

"I have never been there," one of my older students, Pavel, 55 years old, says. "But my son works in Mississippi. He says that the Blacks are taking over."

Everything seems to stop, inside me.

"Research time!" I spout, maybe too loud. "Google it, please. Per capita income, Blacks versus Whites, in the US. Also, four-year college degrees, same question."

Everyone cellphones up.

"College degree is held by 50 percent of Whites," Katya, a youngster with a pug nose and braces, reads. "34 percent of Blacks have one."

"Average Black household has $52,860," Gregor, one student, reads from his cellphone. "The average White household gets $81,060. Per capita income is White people with $36,000 and Black Americans with $23,000 per year."

"Sad news but good research," I say. "Try living in America on $23,000."

Pavel looks uncomfortable.

"Pavel, you said that 'Blacks are taking over'? Taking over what? With what money? With what education?

"You're a European," I say. "But you're starting to sound like some isolated Americans."

75

Kirsti, I Think.

"Kirsti, I may have to end these stories somehow," I say. "But I don't know how."

A gray rainy day finds us outside near the beach again. This makes nine days without sunshine.

During today's four-hour teaching block, I taught Womens' Jiu-Jitsu, Mandarin, French, Spanish, English and Rhumba. When my eyes start closing, the beach tones of gray water and white foam twirl into liquid dripping colors.

"How are you going to end your story here in Odessa?" she asks. "I require novels and dramatic pieces with a definite ending."

"You require too much," I say. "Remember, you're not even a real person. You're a composite character of the New Journalism. From Tom Wolfe, Gay Talese and Pete Hamill."

"But I want a real ending!" she wails.

"Real life doesn't have them," I say. "This war continues. So do these murderous rocket attacks on children. Now, more than ever before."

She pouts.

"I offered to marry you," I continue. "That's a real ending. But-

"When?" she asks. "Never! I don't remember that!"

"But I do," I say.

"Was I asleep?"

"You can't be asleep," I answer. "You're a fictional character. Once, you screamed at me, 'If you write bad private things about me, I will kill you!'"

"Ah, yes," she purrs. "I remember saying that."

"Me, too," I say. "New Journalism does not have nice, clear endings. Neither does anyone's life in Ukraine. I offer you marriage and you threaten me with homicide. What will our readers think about that?"

About the Author

Throughout his early years, Frank Hickey worked the Harlem streets as a private eye. He lived for months undercover in the Cajun bayous of Louisiana, hunting a killer. His adventures have taken him around the world. Living outdoors or in crude huts, he learned local languages and culture.

Carrying the gold shield of a Detective-Investigator/Police Officer, he served in the Manhattan District Attorney's Office on an international serial killer case. Using the Sicilian language, he also went undercover against the New York Mafia.

As a corporal in the Savannah, Georgia Police Department, he learned about the South. Later, as an LAPD Officer, he initiated one-person foot patrols in high-crime areas amid gang wars and drive-by shootings.

37 years of crime adventures from the jungles of Thailand to southern swamps to the streets of New York form his life.

Besides his Max Royster series, Frank also co-wrote Spy, the Movie starring Vincent Pastore [The Sopranos].

Made in United States
Orlando, FL
24 March 2025

59823436R00115